Lecture Notes in Mathematics

Edited by A. Dold and B. Eckmann

604

Banach Spaces of Analytic Functions

Proceedings of the Pelczynski Conference
Held at Kent State University,
July 12-16, 1976

Edited by
J. Baker, C. Cleaver, and J. Diestel

Springer-Verlag
Berlin Heidelberg New York 1977

Editors
J. Baker
C. Cleaver
J. Diestel
Department of Mathematics
Kent State University
Kent, Ohio 44242/USA

AMS Subject Classifications (1970):

ISBN 3-540-08356-1 Springer-Verlag Berlin Heidelberg New York
ISBN 0-387-08356-1 Springer-Verlag New York Heidelberg Berlin

Printed in Germany

Printing and binding: Beltz Offsetdruck, Hemsbach/Bergstr.
2141/3140-543210

TABLE OF CONTENTS

PREFACE

The papers contained in these lecture notes were contributed by participants in an NSF Regional Conference held at Kent State University in the summer of 1976. Several of the speakers at that Conference had already submitted the papers on which their lectures were based to various mathematical journals; we were fortunate to be able to corral the contributions contained herein.

The topic of the Conference was Banach Spaces of Analytic Functions, and the main attraction was the lectures of Professor Aleksander Pelczynski of the Polish Academy of Sciences (who fortuitiously was visiting at Ohio State University from the Fall of 1975 through the summer of 1976). The subject matter covered in these contributions are all related to various aspects and methods discussed in the lectures by Professor Pelczynski. Incidentally, the text of Professor Pelczynski's lectures will appear in monograph form in the Conference Board of Mathematical Sciences series published under the auspices of the American Mathematical Society. People who find things of interest in these notes will find much of interest in Pelczynski's lectures.

As mentioned above, a number of the hour lecturers had submitted their manuscript elsewhere; however, their lectures stimulated great activity in the other Conference participants. This seems as good a place as any to extend our gratitude to those speakers: Tadeusz Figiel, Gdansk; William Johnson, Ohio State University; Robert Kaufman, University of Illinois; Joram Lindenstrauss, Hebrew University; Richard Rochberg, Washington University; Haskell Rosenthal, University of Illinois; Walter Rudin, University of Wisconsin; Allen Shields, University of Michigan; Edgar Stout, University of Washington; Maury Zippin, Hebrew University.

In preparing this volume, we were fortunate to have had expert help from a number of conscientious referees. To these people we extend our thanks: Colin Bennett Sun-Yung Chang, D. J. H. Garling, J. Garnett, D. R. Lewis, R. H. Lohman, J. T. Peck, A. Pelczynski, J. Rainwater, J. R. Retherford, W. Rudin, C. Seifert, J. Shapiro, M. Stoll.

To the Kent State Mathematics Department among of whose members helped greatly with the painful details always attendant to a Conference--particularly, John Fridy, Chuck Seifert, Chuck Giel, Terry Morrison, Arden Ruttan, and Joe Creekmore -- we extend our thanks.

Finally, last but far from least, our thanks to the Departmental secretaries, Julie Froble and Darlene May. They provided calm reason in moments of panic and kept the Conference on an even keel throughout. Further, it's their typing that fills these pages.

AN EXTENSION OF THE RIESZ-THORIN THEOREM

G. Bennett[1]
Indiana University
Bloomington, IN 47401/USA

Abstract. A version of the Riesz-Thorin theorem is given which enables us to interpolate conditions on a matrix A and on its transpose A'. The following result is typical: if both A and A' map ℓ^2 into $\ell^{2/3}$, then A maps ℓ^∞ into ℓ^1.

1. Introduction. By ℓ^p $(0 < p < \infty)$ we denote the space of all complex-valued sequences satisfying $\|x\|_p < \infty$, where

(1) $$\|x\|_p = \begin{cases} (\sum_k |x_k|^p)^{1/p} & \text{if } 0 < p < \infty \\ \sup_k |x_k| & \text{if } p = \infty \end{cases} .$$

We shall be concerned with matrix transformations on these spaces. If A maps ℓ^p into ℓ^q its "norm" is given by

(2) $$\|A\|_{p,q} = \sup_{\|x\|_p \le 1} (\sum_j |\sum_k a_{jk}x_k|^q)^{1/q}.$$

Now suppose that $1 \le p_0, p_1, q_0, q_1 \le \infty$ are given, and that A simultaneously maps ℓ^{p_0} into ℓ^{q_0} and ℓ^{p_1} into ℓ^{q_1}. Then the theorem of the title asserts that A maps ℓ^p into ℓ^q whenever p and q can be expressed in the form

(3) $$\frac{1}{p} = \frac{1-\theta}{p_0} + \frac{\theta}{p_1} \quad \text{and} \quad \frac{1}{q} = \frac{1-\theta}{q_0} + \frac{\theta}{q_1}$$

for some $\theta \in (0, 1)$. Moreover, the norms of A must satisfy

(4) $$\|A\|_{p,q} \le \|A\|_{p_0,q_0}^{1-\theta} \|A\|_{p_1,q_1}^{\theta}.$$

Undoubtedly the "classical values" of the parameters, $1 \le p_0, q_0, p_1, q_1 \le \infty$, considered by Riesz and Thorin, lead to the most important applications. But more general values are not without interest, and the theorem has been extended to the range $0 < p_0, q_0, p_1, q_1 \le \infty$ by Calderon and Zygmund [3], [4]. Their result would seem to be the definitive version of the Riesz-Thorin theorem, any further extension requiring negative values of the parameters. Nevertheless, by adopting a quite natural convention for these values, we shall see that a meaningful extension can indeed be obtained. This extension seems to lie beyond the scope of existing interpolation theories: its proof--and even its statement--being motivated by ideas from the theory of absolutely summing operators. These ideas are discussed first, in section 2, a detailed statement of our main result being deferred to section 3.

2. Motivation. In the theory of absolutely summing operators one often needs

[1]During the preparation of this paper the author was supported in part by National Science Foundation grant MCS76-06906.

estimates of the form $\sum_j (\sum_k |c_{jk}|^r)^{s/r} < \infty$, the estimates to be valid for all matrices C of a given class (say those mapping ℓ^p into ℓ^q, or those admitting a prescribed factorization). Two of the earliest such results, due respectively to Littlewood and Orlicz, are[2]

(5) $$\sum_j (\sum_k |c_{jk}|^2)^{1/2} < \infty \text{ whenever } C: \ell^\infty \to \ell^1$$

and

(6) $$\sum_j \sum_k |c_{jk}|^2 < \infty \qquad \text{whenever } C: \ell^2 \to \ell^1.$$

An extension of these results has been given recently in [1]: if $1 \leq p \leq 2$, then

(7) $$\sum_j (\sum_k |c_{jk}|^2)^{p/2} < \infty \text{ whenever } C: \ell^{p^*} \text{ into } \ell^1$$

(where $p^* = p/(p-1)$ denotes the <u>conjugate</u> of p). It is natural to suspect that (7) is obtainable from (5) and (6) simply by "interpolating". But just the opposite is true: indeed, the <u>proof</u> of (7) leads to new versions of the Riesz-Thorin theorem! The argument given in [1] is indirect and begins with the related estimate:

(8) $$\sum_k (\sum_j |c_{jk}|^2)^{p/2} < \infty \text{ whenever } C: \ell^{p^*} \to \ell^1,$$

itself an easy consequence of Littlewood's inequality. The heart of the matter, then, was to interchange the order of summation in (8). This was achieved by means of the following inequality, valid for <u>arbitrary</u> matrices C:

(9) $$\sum_j (\sum_k |c_{jk}|^2)^{p/2}$$
$$\leq \Big(\sum_j (\sum_k |c_{jk}|^2)^{p/2}\Big)^{2(p-1)/p} \sup_{\|x\|_{p^*} \leq 1} \Big(\sum_j (\sum_k |c_{jk}x_k|^2)^{1/2}\Big)^{2-p}.$$

(The finiteness of the last term, in case $C: \ell^{p^*} \to \ell^1$, again follows easily from Littlewood's result.)

It is inequality (9) with which we shall be concerned in the sequel. Replacing $|c_{jk}|^2$ by a_{jk} it is not difficult to see that (9) is equivalent to

(10) $$\|A\|_{\infty,p/2} \leq \|A'\|_{\infty,p/2}^{2/p^*} \|A\|_{p^*/2,1/2}^{(2-p)/p}$$

(where A' denotes the transpose of A).

Unfortunately, since $p < 2$, the term involving A' cannot be transposed without

---

[2] Results of the second kind lie somewhat deeper. We mention Grothendieck's inequality: $\sum_j (\sum_k |c_{jk}|^2)^{1/2} < \infty$ whenever $C = AB$ with $A: \ell^\infty \to \ell^1$ and $B: \ell^2 \to \ell^\infty$. This includes both the Littlewood and Orlicz inequalities.

losing some information ($\|A'\|_{\infty,p/2} < \infty \Rightarrow \|A\|_{\infty,1} < \infty \Rightarrow \|A'\|_{\infty,1} < \infty \not\Rightarrow \|A'\|_{\infty,p/2} < \infty$). To avoid such losses, let us agree to write

$$\|A\|_{p,q} = \|A\|_{q^*,p^*},$$

<u>no matter what the values of p and q</u>. Of course, both norms make sense, and, indeed, are equal when $p, q \geq 1$. Only the first makes sense when $p > 0 < q < 1$ or $q > 0 < p < 1$, and only the second when $p < 0$, $q \geq 1$ or $q < 0$, $p \geq 1$, or when both are negative. (See the shaded region of the diagram.) If $p < 0 < q < 1$, or if $q < 0 < p < 1$, then neither norm makes sense and we shall have little to say about such cases. In this manner we may consider matrix transformations on certain "negative ℓ^p spaces". Thus, for example, to say that A maps ℓ^{-2} into ℓ^3 means that A' maps $\ell^{3/2}$ into $\ell^{2/3}$.

This convention enables us to express (9) in a very suggestive form. Indeed, setting $\theta = (2 - p)/p$, so that $0 < \theta < 1$, we have

(11) $$\frac{1}{\infty} = \frac{1 - \theta}{(p/2)^*} + \frac{\theta}{p^*/2} \text{ and } \frac{1}{p/2} = \frac{1 - \theta}{1} + \frac{\theta}{1/2}.$$

(10) then becomes

(12) $$\|A\|_{\infty,p/2} \leq \|A\|^{1-\theta}_{(p/2)^*,1} \|A\|^{\theta}_{p^*/2,1/2},$$

and the resemblance with (3), (4) is irrestible!

Unfortunately, we have proved (12) only for matrices with nonnegative entries (recall the substitution $a_{jk} = |c_{jk}|^2$). In the next section we give a version of the Riesz-Thorin theorem which allows this restriction to be removed.

3. <u>The main result</u>. For the statement of our main result it will be convenient to refer to the diagram below. We shall say that a matrix A is <u>bounded at the point</u> (p^{-1}, q^{-1})

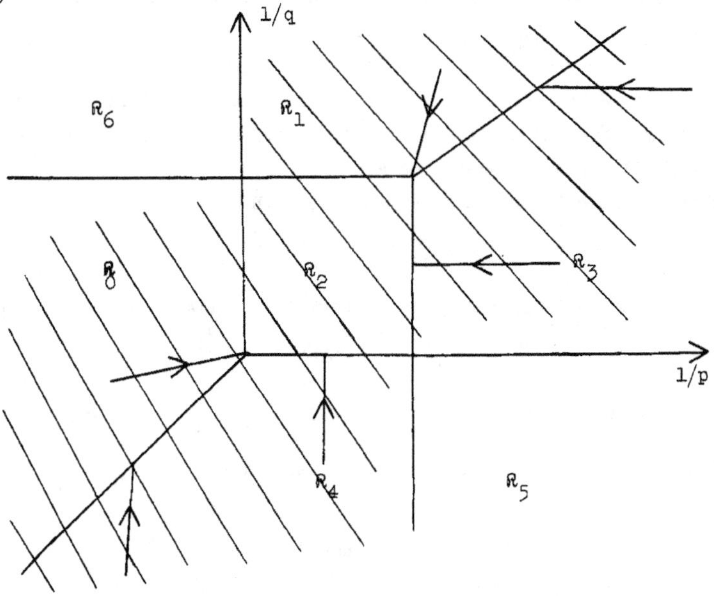

provided that $\|A\|_{p,q} < \infty$ (or, what is equivalent, that A: $\ell^p \to \ell^q$). The convention introduced in section 2 enables us to speak of boundedness at all points of the shaded region: we shall call this region $\mathcal{R}$. We decompose $\mathcal{R}$ into five disjoint regions, as indicated. ($\mathcal{R}_2$ is the closed unit square; the diagonal lines belong to $\mathcal{R}_0$ and $\mathcal{R}_1$; the arrows indicate extrapolation theorems, discussed below.) The results of sections 1 and 2 raise the following question. If three collinear points of $\mathcal{R}$ are given, and a matrix A is bounded at the extremities, must A be bounded at the intermediate point? Certainly, the answer is affirmative in case all three points lie in the positive quadrant (this is the classical interpolation theorem) or in case $a_{jk} \geq 0$ (this is deducible from (12)). Our next result shows that the same assertion holds in general.

Theorem 1. *Suppose a matrix A is bounded at the points $P_0 = (p_0^{-1}, q_0^{-1})$ and $P_1 = (p_1^{-1}, q_1^{-1})$ of $\mathcal{R}$. Then A is bounded at all intermediate points, $P = (p^{-1}, q^{-1})$, of $\mathcal{R}$. Moreover, if $\theta \in (0, 1)$ is chosen so that*

(13) $$\frac{1}{p} = \frac{1-\theta}{p_0} + \frac{\theta}{p_1} \quad \text{and} \quad \frac{1}{q} = \frac{1-\theta}{q_0} + \frac{\theta}{q_1},$$

then we have

(14) $$\|A\|_{p,q} \leq K\|A\|_{p_0,q_0}^{1-\theta} \|A\|_{p_1,q_1}^{\theta},$$

where K is a constant depending only on the line segment.

Proof. Our proof is based on the celebrated technique of Thorin [10], but we also need a deep factorization result of Maurey [5].

It will be convenient to first prove the theorem under the additional hypotheses: $p = \infty$, $0 < q < 1$, $p_0 < 0$, $q_0 = 1$, $p_1 > q_1$, and $0 < q_1 < 1$. (These, in turn, imply that $0 < p_0^* < 1$ and that $q > q_1$.) Given a matrix A we must show that

(15) $$\|A\|_{\infty,q} \leq K\|A'\|_{\infty,p_0^*}^{1-\theta} \|A\|_{p_1,q_1}^{\theta}.$$

We shall assume that A has only finitely many nonzero entries. This involves no loss in generality since the norms (2) clearly satisfy $\|A\| = \sup\|\tilde{A}\|$, the supremum being taken over all finite sub-matrices $\tilde{A}$ of A.

It is then possible to choose a sequence x, with $\|x\|_\infty = 1$, so that

(16) $$\|A\|_{\infty,q} = (\sum_j |\sum_k a_{jk}x_k|^q)^{1/q}.$$

By the converse to Hölder's inequality, the term on the right may be re-written as

(17) $$(\sum_j |\sum_k a_{jk}x_k|^q)^{q_1/q} = \sum_j|\sum_k a_{jk}x_k|^{q_1} Y_j$$

for a suitable nonnegative sequence Y satisfying $\|Y\|_{q/(q-q_1)} = 1$. Applying Maurey's theorem (see section 4), the matrix A admits a factorization: $a_{jk} = b_{jk}w_k$, with $w_k \geq 0$ and

(18) $$\|B'\|_{\infty,1} \|w\|_{-p_0} \leq K\|A'\|_{\infty,p_0^*},$$

where K is a constant depending only on p_0.

Next we consider the function F given by

$$F(z) = \sum_j \Big| \sum_k b_{jk} x_k w_k^{z/\theta} \Big|^{q_1} |Y_j^{(1-z)/(1-\theta)}|.$$

F is clearly bounded and continuous in the strip $0 \le \mathrm{Re}\, z \le 1$. F is also sub-harmonic there, being a (finite) sum of functions of the form $z \to |f(z)|^{q_1}$, with f analytic. The maximum principle for sub-harmonic functions enables us to estimate

(19) $$\sum_j \Big| \sum_k a_{jk} x_k \Big|^{q_1} Y_j = F(\theta)$$

by

$$F(\theta) \le \sup_{-\infty<t<\infty} |F(0 + it)|^{1-\theta} \sup_{-\infty<t<\infty} |F(1 + it)|^{\theta}.$$

Now

$$F(0 + it) = \sum_j \Big| \sum_k b_{jk} x_k \alpha_k \Big|^{q_1} Y_j^{1/(1-\theta)},$$

where α_k is a complex number of modulus one. Applying Hölder's inequality, we obtain

(20) $$\begin{aligned} |F(0 + it)| &\le \Big(\sum_j \Big| \sum_k b_{jk} x_k \alpha_k \Big|\Big)^{q_1} \Big(\sum_j Y_j^{(1-q_1)/(1-\theta)}\Big)^{1/(1-q_1)} \\ &\le \|B\|_{\infty,1}^{q_1} \|x\alpha\|_{\infty}^{q_1} \|Y\|_{q/(q-q_1)}^{q/(q-q_1)(1-q_1)} \\ &\le \|B\|_{\infty,1}^{q_1}. \end{aligned}$$

On the other hand,

(21) $$\begin{aligned} |F(1 + it)| &\le \sum_j \Big| \sum_k b_{jk} x_k w_k^{1/\theta} \alpha_k \Big|^{q_1} \\ &= \sum_j \Big| \sum_k a_{jk} x_k w_k^{(1-\theta)/\theta} \alpha_k \Big|^{q_1} \\ &\le \|A\|_{p_1,q_1}^{q_1} \|w^{-p_0/p_1}\|_{p_1}^{q_1} \\ &= \|A\|_{p_1,q_1}^{q_1} \|w\|_{-p_0}^{-p_0 q_1/p_1}. \end{aligned}$$

It follows from (13), (16), (17), (19), (20) and (21) that

$$\|A\|_{\infty,q} \le \|B\|_{\infty,1}^{1-\theta} \|A\|_{p_1,q_1}^{\theta} \|w\|_{-p_0}^{-p_0\theta/p_1},$$

giving (15) via (18).

We now turn out attention to line segments having one end-point (say P_0) in $\mathcal{R}_0$, the other in $\mathcal{R}_1$. By taking transposes, if necessary, we may assume that the intermediate point $P = (p^{-1}, q^{-1})$, lies in the nonnegative quadrant. Thus, for a given matrix A, we must prove (14) under the hypotheses: (13) and

(22) $$\begin{gathered} p_0 < 0,\ q_0 \le p_0 \ \underline{\text{or}}\ q_0 \ge 1,\ p_1 \ge q_1 \\ 0 < q_1 < 1,\ 0 < p,\ q \le \infty. \end{gathered}$$

Instead of A let us consider the matrix B given by

(23) $$b_{jk} = a_{jk}x_k y_j,$$

where $x \in \ell^p$ and $y \in \ell^{q_0^*}$ are fixed. Upon setting

(24) $$\frac{1}{p_1'} = \frac{1}{p_1} - \frac{1}{p} \text{ and } \frac{1}{q_1'} = \frac{1}{q_1} - \frac{1}{q_0^*},$$

we obtain, by two applications of Hölder's inequality,

(25) $$\|B\|_{p_1',q_1'} \leq \|A\|_{p_1,q_1} \|x\|_p \|y\|_{q_0^*}.$$

Furthermore, by taking transposes, and applying Hölder's inequality again, we have

(26) $$\|B\|_{p_0',1} \leq \|A\|_{p_0,q_0} \|x\|_p \|y\|_{q_0^*},$$

where

(27) $$\frac{1}{p_0'} = \frac{1}{p_0} - \frac{1}{p}.$$

Now it is easy to check, using (13), (22), (24) and (27), that $p_0' < 0$, $p_1' > q_1'$, and $0 < q_1' < 1$. If q' is defined by

(28) $$\frac{1}{q'} = 1 - \theta + \frac{\theta}{q_1'},$$

then we also have $0 < q' < 1$, and so, by (25), (26) and the first part of the theorem

(29) $$\|B\|_{\infty,q'} \leq K\|A\|_{p_0,q_0}^{1-\theta} \|A\|_{p_1,q_1}^{\theta} \|x\|_p \|y\|_{q_0^*},$$

where K is a constant depending only on p_0'. But $x \in \ell^p$ and $y \in \ell^{q_0^*}$ where chosen arbitrarily, so that (29) is equivalent to (14) via (23), (24), (28) and the converse to Hölder's inequality.

Next, we observe that line segments having both end-points in the region $\mathcal{R}_1 \cup \mathcal{R}_2$, or in $\mathcal{R}_0 \cup \mathcal{R}_2$, are covered by applying the classical results either to A or to A'.

Finally, we make two simple observations which enable us to ignore all other line segments. These remarks may be viewed as extrapolation theorems and are perhaps of independent interest. From the monotonicity of the norms (1) it follows that if (p^{-1}, q^{-1}) and (r^{-1}, s^{-1}) are two points of $\mathcal{R}$ with $p \geq r$ and $q \leq s$, then $\|\cdot\|_{r,s} \leq \|\cdot\|_{p,q}$. Thus boundedness at any point implies boundedness at all points of $\mathcal{R}$ lying to the south and to the east. Our second observation is to note that, if $0 < p \leq 1$ and $p \leq q$, then $\|\cdot\|_{r,q} = \|\cdot\|_{p,q}$, where $r = \min(1, q)$. To see this, we denote by e^k the sequence with a one in the kth position and zeroes elsewhere. Then if A maps ℓ^p into ℓ^q, we have, by the r-convexity of the norm $\|\cdot\|_q$,

$$\|Ax\|_q = \|\sum_k x_k(Ae^k)\|_q$$
$$\leq \|x\|_r \sup_k \|Ae^k\|$$
$$\leq \|x\|_r \|A\|_{p,q}.$$

These remarks enable us to ignore any line segment, P_0P_1, having an end-point in the region $\mathcal{R}_3 = \{(p^{-1}, q^{-1}): \; 0 < p < 1 \text{ and } p < q\}$. For if such a line segment is given, and a matrix A is assumed bounded at P_0 and P_1, then our second observation enables us to extrapolate the offending end-point condition(s) westward to the boundary of $\mathcal{R}_3$. The resulting intermediate estimate on A is then stronger, courtesy of the first observation, than the estimate at P proposed by the theorem. Similar remarks may be applied to the region $\mathcal{R}_4$, and the theorem is proved.

4. The factorization theorem. In this section we discuss the factorization result that was used in the proof of Theorem 1. Let $0 < p < 1$ be given, and suppose that A maps ℓ^∞ into ℓ^p: we must show that A admits a factorization, $A = DB$, where B maps ℓ^∞ into ℓ^1 and where D is a diagonal matrix which maps ℓ^1 into ℓ^p. This result in fact is known, being equivalent, by Theorem 8 of [5], to the assertion that every nuclear operator on ℓ^1 be p-absolutely summing. Unfortunately, however, this last result (see [5], Theorem 93) lies very deep, and its proof requires an extensive knowledge of the theory of operator ideals. For the benefit of the reader who is unfamiliar with this area we give here a self-contained presentation of the factorization result. The essential ingredient is a magnificant "sliding humps" argument used by Rosenthal in [8].

We begin with a lemma of some interest in its own right. Suppose a sequence of matrices $A^{(1)}, A^{(2)}, \cdots$ is given, $A^{(i)}$ having m_i rows. If A denotes the matrix with $(\sum_i m_i)$ rows, formed by juxtaposing $A^{(1)}, A^{(2)}, \cdots$ in the obvious fashion, then clearly $\|A\|_{\infty,1} \geq \sup_i \|A^{(i)}\|_{\infty,1}$. The lemma gives a more precise estimate.

Lemma 1. With the above notations we have

$$\sum_i \|A^{(i)}\|_{\infty,1}^2 \leq 2\|A\|_{\infty,1}^2.$$

Proof. The proof is much simpler if we consider horizontal juxtapositions, $A = (A^{(1)}|A^{(2)}| \cdots)$, instead of vertical ones. That we may do this is easily seen by taking transposes.

Choosing unit vectors $x^{(i)} \in \ell^\infty$ (of the appropriate dimension) so that $\|A^{(i)}x^{(i)}\|_1 = \|A^{(i)}\|_{\infty,1}$, we obtain, by Minkowski's inequality,

$$\begin{aligned}
&(\sum_i \|A^{(i)}\|_{\infty,1}^2)^{1/2} \\
&= (\sum_i (\sum_j |\sum_k a_{jk}^{(i)} x_k^{(i)}|)^2)^{1/2} \\
&\leq \sum_j (\sum_i |\sum_k a_{jk}^{(i)} x_k^{(i)}|^2)^{1/2} .
\end{aligned}$$

Introducing the Rademacher functions $r_1, r_2, \cdots$, it follows from Khinchine's inequality that

$$\sum_j (\sum_i |\sum_k a_{jk}^{(i)} \; x_k^{(i)}|^2)^{1/2}$$

$$\le C\sum_j \int_0^1 \Big|\sum_i\sum_k a_{jk}^{(i)}\, x_k^{(i)} r_i(t)\Big|\,dt$$

$$\le C \operatorname*{ess.\,sup}_{t\in[0,1]} \sum_j \Big|\sum_i\sum_k a_{jk}^{(i)} x_k^{(i)}\, r_i(t)\Big|$$

$$\le C\|A\|_{\infty,1}\, \sup_i \|x^{(i)}\|_\infty .$$

This gives the lemma with C^2 in place of 2. (The actual value of the constant is of no importance to us here, but the reader interested in taking $C^2 = 2$ should consult remark (4) of section 5.)

We are now in a position to prove the factorization result. It will be convenient to restrict attention to matrices with only finitely many nonzero entries (finite matrices), for then the various "sups" and "infs" are actually attained. The extension to arbitrary matrices is routine, and, besides, we have used only the finite version in section 3.

Theorem 2. Let $0 < p < 1$. Then for all (finite) matrices A we have

$$\inf\{\|d\|_{p/)1-p)}\|B\|_{\infty,1}\colon\ A = (\text{diag}d)\cdot B\} \le K\|A\|_{\infty,p},$$

where K is a constant depending only on p.

Proof. Given $0 < \alpha < 1$ and a positive integer n, we choose β, γ, $\delta > 0$, in that order, to satisfy

(30) $$\delta\Big(1 - \Big(\frac{n}{\beta\gamma^p}\Big)^{(1-p)/p}\Big)(1-\beta)^{(1-p)/p} - \gamma^{1-p} \ge \alpha\delta.$$

Now if the theorem is false, it is possible to find a matrix A with

(31) $$\|A\|_{\infty,p} \le 1$$

and such that

(32) $$\min\{\|d\|_{p/(1-p)}\|B\|_{\infty,1}\colon\ A = (\text{diag}d)\cdot B\} = \delta.$$

Thus A may be expressed in the form: $a_{jk} = d_j b_{jk}$, where

(33) $$\|B\|_{\infty,1} = \delta$$

and

(34) $$\|d\|_{p/(1-p)} = 1.$$

Next we construct inductively disjoint sets $J_1, \cdots, J_n$ of positive integers, and unit vectors $x^{(1)}, \cdots, x^{(n)} \in \ell^\infty$ to satisfy the following three conditions.

(35) $$\Big|\sum_k a_{jk}x_k^{(i)}\Big| > \gamma|d_j|^{1/(1-p)} \quad \text{for } j \in J_i$$

(36) $$\sum_{j\in J_i} |d_j|^{p/(1-p)} \le \gamma^{-p}$$

(37) $$\sum_{j\in J_i} \Big|\sum_k b_{jk}x_k^{(i)}\Big| \ge \alpha\delta.$$

Condition (37) is the most important one: the other two are needed only to push through the induction argument.

Using (33), we choose $x^{(1)}$ so that $\|Bx^{(1)}\|_1 = \delta$, and then set $J_1 = \{j\colon (35)$ holds for $i = 1\}$. It follows from (31) and (30) that (36) and (37) also hold for J_1. Now suppose that $x^{(1)}, \cdots, x^{(m-1)}$ and $J_1, \cdots, J_{m-1}$ have been determined $(1 < m \leq n)$. Letting $J = \bigcup_{i=1}^{m-1} J_i$, we define a sequence c by setting

$$c_j^{p/(1-p)} = \begin{cases} \frac{\beta}{n} |\sum_k a_{jk} x_k^{(i)}|^p & \text{if } j \in J_i \text{ for some } i < m \\ (1 - \sum_{r \in J} c_r^{p/(1-p)}) \dfrac{d_j^{p/(1-p)}}{\sum_{s \in J^c} d_s^{p/(1-p)}} & \text{if } j \in J^c \end{cases}$$

It is clear, using (31), that $\|c\|_{p/(1-p)} = 1$; moreover, it follows from (35) that $c_j = 0$ only when $a_{jk} = 0$ for all values of k. Thus, by (32), we may choose $x^{(m)}$, with $\|x^{(m)}\|_\infty = 1$, in such a way that

$$(39) \qquad \sum_{j \in J} |\sum_k c_j^{-1} a_{jk} x_k^{(m)}| \leq (\frac{n}{\beta\gamma^p})^{(1-p)/p} \delta$$

and then from (38) and (39) that

$$(40) \qquad \sum_{j \in J^c} |\sum_k b_{jk} x_k^{(m)}| \geq \delta(1 - (\frac{n}{\beta\gamma^p})^{(1-p)/p})(1-\beta)^{(1-p)/p}.$$

Setting

$$J_m = \{j\colon |\sum_k b_{jk} x_k^{(m)}| > \gamma |d_j|^{p/(1-p)}\} \cap J^c,$$

it is clear that (35) and (36) hold when $i = m$. To see that (37) is also satisfied we note that

$$\sum_{j \in J^c \cap J_m^c} |\sum_k b_{jk} x_k^{(m)}| \leq \gamma^{1-p},$$

and apply (40) and (30). This completes the induction.

To obtain the required contradiction we apply the lemma to the matrix B. Recalling (33) and (37), this gives $n\alpha^2 \delta^2 \leq 2\delta^2$, which is impossible since α and n were chosen artibtarily.

5. Closing remarks. (1) Throughout this paper we have restricted attention to the ℓ^p spaces. The techniques used to prove Theorem 1, however, are perfectly general, and analogous extensions of the Riesz-Thorin theorem can be formulated for other L^p spaces.

(2) We have seen, in Lemma 1, that "stacking" matrices leads to an appreciable increase in $\|\cdot\|_{\infty,1}$. Other norms behave quite differently: for example, on Hilbert space, it is possible to stack infinitely many matrices without increasing $\|\cdot\|_{2,2}$ at all. This raises the question of whether the exponent "2" appearing in Lemma 1

is the best (i.e. smallest) possible. An affirmative answer is given by the following general observation, whose proof is omitted. <u>With the notation of Lemma 1, we have $\sum_i \|A^{(i)}\|^r_{p,q} \leq C\|A\|^r_{p,q}$ (for some constant C depending only on p and q) if and only if</u>

$$r \geq \begin{cases} \frac{q \min(2,p)}{\min(2,p)-q} & \text{when } 1 \leq q < \min(2,p) \\ \infty & \text{when } q \geq \min(2,p) \geq 1. \end{cases}$$

(3) We have considered only complex-valued sequences in this paper, but it is readily seen, by using the elementary estimates: $\|A\|^{\mathbb{R}}_{p,q} \leq \|A\|^{C}_{p,q} \leq \sqrt{2}\,\|A\|^{\mathbb{R}}_{p,q}$, that all our results are valid in the real case too. In general, however, we have $K^{\mathbb{R}} \neq K^{C}$ for the constant appearing in Theorem 1. Riesz himself observed that $K^{\mathbb{R}}$ is necessarily greater than one for certain line segments: on the other hand it is conceivable that we can always take $K^C = 1$. (It is possible to show that a <u>uniform</u> constant K exists for all line segments.)

(4) The best value for the constant C appearing in the proof of Lemma 1 is known to be $\sqrt{2}$. This fact is quite difficult to establish and was obtained only recently by Szareck [9]. Taking $A^{(1)}$ and $A^{(2)}$ to be the 1×2 matrices (1, 1) and (1, -1), we see that the constant "2" of Lemma 1 is best possible for <u>real</u> scalars. I do not know whether the same is true in the complex case. We remark, however, that in the latter case, the argument of Szareck can be avoided, as follows. Instead of introducing the Rademacher functions we consider a sequence $f_1, f_2, \cdots$ of independent random variables taking the three values: 1, $e^{2\pi i/3}$ and $e^{4\pi i/3}$. We then have $E(|\sum_i x_i f_i|^4) = 2(\sum_i |x_i|^2)^2$, which suffices for a Khinchine-type inequality with constant $C = \sqrt{2}$. (If the constant "2" of Lemma 1 could be reduced, we could then take $n = 2$ in the proof of Theorem 2, and thereby eliminate the induction argument.)

(5) It is natural to ask whether we can extend the notion of boundedness frm $\mathbb{R}$ to the entire complex plane in such a way that the Riesz-Thorin theorem remains valid. For the region $\mathbb{R}_5$ this is easy: we simply take boundedness at any point of $\mathbb{R}_5$ to mean $\|\cdot\|_{1,\infty} < \infty$. On the other hand, I see no way of handling $\mathbb{R}_6$.

(6) Given a matrix A it is interesting to study the set $\mathbb{R}(A) = \{(p^{-1},q^{-1}): \|A\|_{p,q} < \infty\}$. For example, the Riesz-Thorin theorem asserts that $\mathbb{R}(A)$ is a convex subset of $\mathbb{R}$. Now a simple cardinality argument shows that not every convex subset of $\mathbb{R}$--even one having the appropriate extrapolation properties--can be realized as an $\mathbb{R}(A)$. In the original draft of this paper we proved the following result, which may be viewed as a converse to the Riesz-Thorin theorem. <u>The subsets of $\mathbb{R}$ expressible in the form $\mathbb{R}(A)$ are precisely those which are convex, F_σ, and closed under extrapolations</u>. My brother, Colin, however, has brought to my attention the paper [6], where the same result is established for operators on the spaces $L^p[0,1]$.

Some simplifications in the argument of Riemenschneider are possible (especially on pages 408-415 of [6]), but these differences do not merit publication.

(7) Finally, we remark that Theorem 1 is not trivial even when the matrix entries are assumed to be nonnegative. (This is of interest since the classical versions of the theorem then become routine exercises involving Hölder's inequality.) A proof of this special case (with constant K = 1) can be obtained from Lemma 7.3 of [2]. This result, which generalizes (9), constitutes a very satisfactory substitute for Minkowski's inequality when handling expressions of the form $\sum_j (\sum_k |c_{jk}|^r)^{s/r}$. It was Lemma 7.3, rather than (9), which led us to suspect that a general principle was at work here.

REFERENCES

1. G. Bennett, Inclusion mappings between ℓ^p spaces, Jour. Funct. Anal. 12 (1973), 420-427.
2. G. Bennett, Hadamard multipliers, to appear.
3. A. P. Calderon and A. Zygmund, On the theorem of Hausdorff-Young, Ann. of Math. Studies 25, Princeton University Press (1950).
4. A. P. Calderon and A. Zygmund, A note on the interpolation of linear operations, Studia Math. 12 (1951), 194-204.
5. B. Maurey, Théorèmes de factorisation pour les opérateurs linéaires à valeurs dans les espaces L^p, Astérisque 11, Soc. Math. France (1974).
6. S. D. Riemenschneider, The L-characteristics of linear operators on $L^{1/\alpha}[0,1]$, Jour. Funct. Anal. 8 (1971), 405-421.
7. M. Riesz, Sur les maxima des formes bilinéaires et sur les functionelles linéaires, Acta Math. 49 (1926), 465-497.
8. H. P. Rosenthal, On subspaces of L^p, Ann. of Math. 97 (1973), 344-373.
9. St. J. Szareck, On the best constant in the Khinchine inequality, to appear.
10. G. O. Thorin, An extension of a convexity theorem due to M. Riesz, Kungl. Fysiografiska Sallskapets i Lund Fordhandlinger 8 (1939), No. 14.

SOME ALGEBRAS OF BOUNDED ANALYTIC FUNCTIONS CONTAINING THE DISK ALGEBRA

S. Y. Chang(*)
The Institute for Advanced Study
Princeton, NJ 08540/USA

D. E. Marshall(**)
University of California
Los Angeles, CA 91403/USA

1. Introduction. Let Δ be the open unit disk in the complex plane and let H^∞ be the uniformly closed subalgebra of $L^\infty(d\theta/2\pi, \partial\Delta)$ consisting of L^∞ functions whose negative Fourier coefficients vanish. Equivalently, $f \in H^\infty$ if and only if f is the boundary-value function of a bounded analytic function on Δ, which we will also call f. If B is any closed algebra lying between H^∞ and L^∞, we let C_B be the C^*-algebra generated by invertible Blaschke products in B. Equivalently, C_B is the uniform closure of the set of functions of the form $f = \bar{b}_0 \sum_1^n \lambda_i b_i$ where $\lambda_i \in \mathbb{C}$ and b_i are Blaschke products in $H^\infty \cap B^{-1}$. In [6] and [22] it is shown that H^∞ and C_B generate B. Moreover, Chang [7] has shown that the linear span $H^\infty + C_B$ equals B[1]. We wish to consider the algebras $H^\infty \cap C_B$ as generalizations of both the disk algebra, A, and H^∞. Indeed, if $B = L^\infty$ then Douglas and Rudin [11] have shown that $C_B = L^\infty$, so that $H^\infty \cap C_B = H^\infty$. It is shown in [19] that the algebra generated by H^∞ and $\bar{z}$ is the smallest closed subalgebra of L^∞ properly containing H^∞. In this case the corresponding C_B algebra is called C, the set of continuous functions on $\partial\Delta$, and $H^\infty \cap C_B = A$. Several theorems will be proved to show that each algebra $H^\infty \cap C_B$ has many of the properties of both A and H^∞.

In [27], D. Sarason commented that the Douglas property [6], [22], and Wermer's maximality theory [28] could be stated in a similar fashion. Moreover the solution of the Douglas problem was analogous to the proof of Wermer's theorem given by Hoffman and Singer [18, p. 93]. In section 3 we will prove a theorem about algebras between $H^\infty \cap C_B$ and C_B that contains both Wermer's theorem and tl solution to the Douglas problem as special cases. The proof uses a suggestion of N. Jewell [17] to reorder the lemmas in both [6] and [22] to prove the Douglas problem without the use of the corona theorem.

In section 4 we prove that the closed unit ball of $H^\infty \cap C_B$ is the closure of the convex hull of the Blaschke products in $H^\infty \cap C_B$. This contains as special cases the theorems of Fisher [12] for A and Marshall [21] for H^∞. J. Wermer has

---

[1] We will give a modified proof of this in section 2.

(*) Research support in part by NSF grant MPS 7506675.

(**) Research support in part by NSF-NATO postdoctoral fellowship in Science.

asked which algebras between A and H^∞ have the corona property. Dawson [9] has shown that the algebra generated by z and $(1-z)^i$ does not have the corona property. In section 4 we prove that the disk is dense in the maximal ideal space of $H^\infty \cap C_B$. This was proved in the case $C_B = L_E^\infty$ by Detraz [10] and by Gamelin and Garnett [15]. Here $E \subset \partial\Delta$ and L_E^∞ is the set of functions in L^∞ which are continuous on E. Finally we use the main theorems of this paper to prove a theorem on the structure of algebras between $H^\infty \cap C_B$ and C_B.

In the following, if S is a subset of L^∞, [S] will denote the closed algebra generated by S. If S is a subalgebra of L^∞, $\mathfrak{M}(S)$ will denote its maximal ideal space and S^{-1} will denote the invertible elements of S. If $f \in L^\infty$ then d(f, S) will denote the distance from f to the set S. That is, $d(f, S) = \inf\{\|f - s\|_\infty : s \in S\}$. We will let B always denote an arbitrary, but fixed, closed algebra lying between H^∞ and L^∞, and C_B will be the corresponding C^*-algebra as described above.

It is a pleasure to acknowledge helpful discussions with J. Garnett.

2. <u>A Variant of Nevanlinna's Theorem</u>. In this section we will reprove results in [7] which will be used many times in the following sections. The proofs below are given for several reasons. They can be generalized as in section 5 to algebras between $H^\infty \cap C_B$ and C_B, they yield an improved distance estimate, and they are more elementary.

We begin with a theorem of D. Sarason (see [7]), originally proved using Toeplitz operators.

<u>Theorem 2.1</u>. Suppose v is a unimodular function in L^∞ with $d(v, H^\infty) = 1$ and $d(v, [H^\infty, \bar{z}]) < 1$. Then $\bar{v} \in [H^\infty, v]$.

<u>Proof</u>. Since $\bar{z} \in [H^\infty, v]$, without loss of generality we can assume $d(v, H^\infty) = 1$ and $\|zv - h\| < 1$ for some $h \in H^\infty$. We claim h is invertible in H^∞. To see this, suppose $|h|$ is not bounded below on Δ. Replacing h by $h - h(z_0)$ with $|h(z_0)|$ small, we can assume

$$\|zv - h_1 \quad (z - z_0)/(1 - \bar{z}_0 z)\| < 1$$

for some $h_1 \in H^\infty$ and some $z_0 \in \Delta$. Thus

$$\|1 - \bar{v}h_1 |1 - z_0\bar{z}|^2/(1 - \bar{z}_0 z)^2\| < 1,$$

so that

$$\|v - h_1(1 - |z_0|)^2/(1 - \bar{z}_0 z)^2\| < 1.$$

But this contradicts $d(v, H^\infty) = 1$, so we must have $h^{-1} \in H^\infty$. Now $\|1 - \overline{zvh}\| < 1$ so there exists $\alpha > 0$ such that $\|1 - zvh^{-1}\alpha\| < 1$. Thus

$$\bar{v} = zh^{-1}\alpha(zvh^{-1}\alpha)^{-1} \in [H^\infty, v].$$

In [1] a version of Nevanlinna's theorem [24, Satz 7] is proved. It may be stated as follows. If $f \in L^\infty$, with $d(f, H^\infty) < 1$, then there exists a unimodular function $u \in L^\infty$ such that $u \in f + H^\infty$ and $d(u, zH^\infty) = 1$. Chang [7] deduced the next corollary from this version of Nevanlinna's theorem and Theorem 2.1. We include the proof for the convenience of the reader.

Corollary 2.2. If $g \in H^\infty$ and if b is an inner function in $H^\infty \cap C_B$ with $d(g\bar{b}, H^\infty) < 1$, then there exists a unimodular function $u \in g\bar{b} + H^\infty$ with $u \in C_B$. Furthermore, if $g \in H^\infty \cap C_B$ then $u \in g\bar{b} + H^\infty \cap C_B$.

Proof. By Nevanlinna's theorem there exists a unimodular function $u = g\bar{b} + h$ for some $h \in H^\infty$, with $d(u, zH^\infty) = 1$. If $v = \bar{z}u$, then $d(v, H^\infty) = 1$ and $d(v, [H^\infty, \bar{z}]) \leq d(u, H^\infty) = d(g\bar{b}, H^\infty) < 1$. So by Theorem 2.1, $\bar{u} \in [H^\infty, g\bar{b}]$. Thus bu is an inner function and $\overline{bu} \in [H^\infty, \bar{b}] \subset B$. Hence $bu \in C_B$ and therefore, $u \in C_B$. If $g \in H^\infty \cap C_B$ then clearly $h \in H^\infty \cap C_B$.

Remark: If b is analytic across an arc $I \subset \partial\Delta$ then bu is an inner function which is also analytic across I.

To see this, note that if $\varphi \in \mathfrak{M}(H^\infty)$ and $\varphi(z) \in I$ then $|\varphi(b)| = 1$. This says that $\varphi \in \mathfrak{M}[H^\infty, \bar{b}]$, so that $|\varphi(bu)| = 1$. Thus bu does not vanish on any of the fibers of $\mathfrak{M}(H^\infty)$ lying over I and hence must be analytic across I.

That $H^\infty + C_B$ is closed is implied by the following lemma (see [8]).

Lemma 2.3. If $f \in C_B$ then $d(f, H^\infty \cap C_B) = d(f, H^\infty)$.

We remark that if $B = [H^\infty, \bar{z}]$ this gives the well-known fact that $d(f, A) = d(f, H^\infty)$ for all $f \in C$. For the case $B = [H^\infty, L_E^\infty]$, this lemma was proved in [8].

Proof. Clearly $d(f, H^\infty \cap C_B) \geq d(f, H^\infty)$. To prove the opposite inequality, we first consider functions of the form $f = g\bar{b}$ where g and b are in $H^\infty \cap C_B$ and b is a Blaschke product. By Corollary 2.2, if $d(g\bar{b}, H^\infty) < 1$, then $d(g\bar{b}, H^\infty \cap C_B) \leq 1$. Thus $d(g\bar{b}, H^\infty) = d(g\bar{b}, H^\infty \cap C_B)$. For general $f \in C_B$, we note that f may be approximated by functions of the form $\bar{b}_0 \, \Sigma \, \lambda_i b_i$ where the b_i are Blaschke products in C_B and the $\lambda_i \in \mathbb{C}$. The lemma follows.

Finally, we deduce the following result in Chang [7].

Theorem 2.4. $B = H^\infty + C_B$. In other words, the linear span $\{h + k\colon h \in H^\infty, k \in C_B\}$ is closed, is an algebra, and is equal to B.

Proof. By the Douglas property for algebras between H^∞ and L^∞ ([6], [22]), the set of functions $S = \{g\bar{b}\colon g \in H^\infty,\ b \text{ is a Blaschke product in } C_B\}$ is dense in B. By Corollary 2.2, if $g\bar{b} \in S$ then $g\bar{b} = h + u\cdot 2\|g\|_\infty$ for some $h \in H^\infty$ and $u \in C_B$. Since $H^\infty + C_B$ is closed, $B = H^\infty + C_B$ proving the theorem.

We remark that many of the results of this paper would be considerably easier if a more general version of Nevanlinna's theorem were true. That is, if $f \in C_B$, $\|f\|_\infty < 1$, then there does not necessarily exist a unimodular function in

$f + H^\infty \cap C_B$. Examples where it fails for the case $H^\infty \cap C_B = A$, the disk algebra, are given in [3, p. 228]. Corollary 2.2, however, says this version is true for f in a dense subset of C_B.

3. A Generalization of Wermer's Maximality Theorem. It is well known that the disk algebra A is a Dirichlet subalgebra of C, the continuous functions on $\partial\Delta$. Also H^∞ is a strongly logmodular subalgebra of L^∞. In particular, A and H^∞ are logmodular subalgebras of C and L^∞ respectively.

Lemma 3.1. $H^\infty \cap C_B$ is a logmodular subalgebra of C_B.

Proof. Let u be a real-valued function in C_B. Then there exist Blaschke products $b_0, b_1, \cdots, b_n$ in C_B such that $\|e^u - \bar{b}_0 \Sigma \lambda_i b_i\| < \epsilon$. Let $\Sigma \lambda_i b_i = wf$ where w is an inner function and f is a outer function in H^∞. We claim $w \in C_B$ or in other words, $\bar{w} \in B$. Note that $\bar{f} = \bar{w}\overline{wf} \in B$. Since $|f| \geq \exp(\inf u) - \epsilon$ on $\partial\Delta$, $f^{-1} \in H^\infty$ if ϵ is sufficiently small. It follows that $\bar{f} \neq 0$ on $\mathfrak{M}(B)$, since $\mathfrak{M}(B)$ can be viewed as a subset of $\mathfrak{M}(H^\infty)$ by the logmodularity of H^∞ in L^∞. Thus $\bar{f}^{-1} \in B$ and we see that $\bar{w} = \overline{wf}\cdot\bar{f}^{-1} \in B$. This proves the claim. Since C_B is a C^*-algebra contained in B, every element of $\mathfrak{M}(C_B)$ extends to an element of $\mathfrak{M}(B)$ [20, Theorem 9.5.3] and, therefore, to an element of $\mathfrak{M}(H^\infty)$. So $f = \bar{w}\cdot wf \in C_B$ and since $f \neq 0$ on $\mathfrak{M}(H^\infty)$, we conclude $f \neq 0$ on $\mathfrak{M}(C_B)$ and hence $f^{-1} \in C_B$. Thus f is an invertible function in $H^\infty \cap C_B$ and $\log|f|$ approximates u.

Theorem 3.2. Suppose D is a closed algebra such that $H^\infty \cap C_B \subset D \subset C_B$. Then D is a Douglas algebra over $H^\infty \cap C_B$. That is, D is generated by $H^\infty \cap C_B$ and the inverses of the invertible Blaschke products in D.

We remark that for $B = L^\infty$, this gives the solution to the Douglas problem. For $B = [H^\infty, \bar{z}]$ this contains Wermer's maximality theorem. That is, if D is a closed algebra between A and C, $D \neq A$, then D contains the complex conjugate of a finite Blaschke product, $\bar{b}$, by Theorem 3.2. So for some Blaschke product $b_1 \in H^\infty \cap C_B$ and some $a \in \Delta$, $b_1\bar{b} = (\bar{z} - \bar{a})/(1 - a\bar{z}) \in D$ and hence $\bar{z} = (b_1\bar{b} + \bar{a})/(1 + ab_1\bar{b})$ is in D. By the Stone-Weierstrass theorem D = C.

Proof. It suffices to consider algebras of the form $D = [H^\infty \cap C_B, f]$ for some $f \in C_B$. Furthermore we can assume $f^{-1} \in D$ and $\frac{1}{2} < |f| < 1$ on $\partial\Delta$. Since $H^\infty \cap C_B$ is logmodular, for $\beta < 1$, there exists a $g_\beta \in (H^\infty \cap C_B)^{-1}$ such that $0 \leq |g_\beta| - |f| \leq (1 - \beta)/4$. When $\frac{1}{2} < \beta < 1$, this implies $\beta < (1 + \beta)/2 < |fg_\beta^{-1}| \leq 1$. Let $u_\beta = fg_\beta^{-1}$ and note that $[H^\infty \cap C_B, f] = [H^\infty \cap C_B, u_\beta, u_\beta^{-1}]$ for each β, $\frac{1}{2} < \beta < 1$. These u_β will replace the unimodular function u in [22].

For each α, $\frac{1}{2} < \alpha < 1$, find $\beta = \beta(\alpha) < 1$ as in Lemmas 1 and 2 of [22]. Here we work with an arbitrary bounded harmonic function u, $\|u\|_\infty \leq 1$. We apply the construction in [22] to the harmonic extension of u_β to Δ. This construction is based on the construction in [4] and modifies an argument in [29].

Briefly, we surround the places in Δ where $|u_\beta| \leq \alpha$ with contours Γ contained in $\{z: |u_\beta| \leq \beta\}$. The contours Γ induce a Carleson measure. That is, if S is a sector of the form $\{re^{i\theta}: 1-\delta < r < 1, \theta_0 \leq \theta \leq \theta_0 + \delta\}$ then $|\Gamma \cap S|$ denotes the length of $\Gamma \cap S$ and C is a universal constant. We place points $\{a_n\}$ on Γ, evenly spaced in the pseudo-hyperbolic metric. Since Γ induces a Carleson measure, the Blaschke product with zero sequence $\{a_n\}$ will be an interpolating Blaschke product ([5], see also [16]), call it b_α. The only place in [22] where it was required that the harmonic function u satisfy $|u| = 1$ a.e. on $\partial\Delta$ was to prove $|\Gamma \cap \partial\Delta| = 0$. But $|u_\beta| > \beta$ a.e. on $\partial\Delta$ and $\Gamma \subset \{z: |u_\beta| \leq \beta\}$, so we conclude $|\Gamma \cap \partial\Delta| = 0$. Thus b_α has the following properties in Δ:

(3.1) If $b_\alpha(z) = 0$, then $|u_\beta(z)| \leq \beta$.

(3.2) If $|u_\beta(z)| \leq \alpha$, then $|b_\alpha(z)| \leq 1/10$.

To prove Theorem 3.2 we will show that

$$[H^\infty \cap C_B, f] = [H^\infty \cap C_B, \{\overline{b}_\alpha\}\ \tfrac{1}{2} < \alpha < 1].$$

First we show each $b_\alpha \in C_B$. If $\varphi \in \mathfrak{M}([H^\infty, f])$ and $\varphi(b_\alpha) = 0$ then φ is in the closure of the zeros of b_α. So by (3.1), $|\varphi(u_\beta)| \leq \beta$. But $\varphi(u_\beta)\varphi(u_\beta^{-1}) = 1$, so $|\varphi(u_\beta)| \geq 1/\|u_\beta^{-1}\| \geq (1 + \beta)/2 > \beta$. This contradiction shows $b_\alpha \neq 0$ on $\mathfrak{M}([H^\infty, f])$. Hence $\overline{b}_\alpha \in [H^\infty, f] \subset B$. That is $b_\alpha \in C_B$.

Let $A_1 = [H^\infty \cap C_B, \{\overline{b}_\alpha\}\ \frac{1}{2} < \alpha < 1]$. To see $A_1 \subset [H^\infty \cap C_B, f]$, we now only need to show $b_\alpha \neq 0$ on $\mathfrak{M}([H^\infty \cap C_B, f])$. Since $H^\infty \cap C_B$ is a logmodular subalgebra of C_B, each $\varphi \in \mathfrak{M}(H^\infty \cap C_B)$ has a unique extension to a bounded linear functional (with the same norm) on C_B, which we also call φ. Thus we view $\mathfrak{M}([H^\infty \cap C_B, f])$ as a subset of $\mathfrak{M}(H^\infty \cap C_B)$. By Corollary 2.2, if $g \in H^\infty$ there is a function $u \in H^\infty \cap C_B$ with $u = g + b_\alpha h$ for some $h \in H^\infty$. Hence if S if the zero sequence for the interpolating Blaschke product b_α, then $H^\infty \cap C_B|_S = H^\infty|_S \cong \ell^\infty(S)$. So by [18, p. 205], if $\varphi \in \mathfrak{M}([H^\infty \cap C_B, f[)$ and $\varphi(b_\alpha) = 0$ then φ is in the closure of the zeros of b_α in $\mathfrak{M}[H^\infty \cap C_B]$. By exactly the same argument given above we obtain a contradiction. So $\overline{b}_\alpha \in [H^\infty \cap C_B, f]$ for each α, $\frac{1}{2} < \alpha < 1$.

Conversely we wish to show $f \in A_1$. To see this, for each fixed α, we let G^α be the region $G^\alpha = \{z \in \Delta: |b_\alpha(z)| > 1/10\}$. Applying Lemma 2 and Lemma 5 in [6] to the region G^α and the function u_β satisfying (3.2) above, one can show that the function u_β satisfies the following property: (with notations as in [6]).

The measure μ^α defined by $d\mu^\alpha = \chi_{G^\alpha}(re^{i\theta})(1 - r)|\nabla u_\beta(re^{i\theta})|^2 r\, dr\, d\theta$ is a Carleson measure on the unit disk with the Carleson constant $\leq C(1 - \alpha)$, where C is an universal constant. (i.e., for each sector of the form $S = \{re^{i\theta}: 1-\delta<r<1, \theta_0 \leq \theta \leq \theta_0 + \delta\}$ we have $\mu^\alpha(S) \leq C(1 - \alpha)\delta$.)

We can then apply Theorem 6 in [6] to the functions b_α and u_β and obtain:

$$(3.3)\qquad \limsup_{n\to\infty}\ d(u_\beta\, b_\alpha^n\, H^\infty) \le C\,(1-\alpha)^{1/2}.$$

Now $u_\beta b_\alpha^n \in C_B$, so by Lemma 2.3 and (3.3), for n sufficiently large we have

$$\begin{aligned} d(f, A_1) = d(g_\beta u_\beta, A_1) &\le \|g_\beta\|_\infty d(u_\beta b_\alpha^n, H^\infty \cap C_B) \\ &\le 2d(u_\beta b_\alpha^n, H^\infty) \\ &\le 2C(1-\alpha)^{1/2}. \end{aligned}$$

Letting α approach 1, we see that $f \in A_1$.

<u>Remark</u>: In the proof above, we have used an observation by N. Jewell [17] which combines the arguments in [6] and [2] to settle the Douglas problem without using the corona theorem.

4. <u>Inner Functions and the Corona Property for $H^\infty \cap C_B$</u>. The following theorem for the case $H^\infty \cap C_B = A$ was proved by Fisher [12] and for the case $H^\infty \cap C_B = H^\infty$ by Marshall [21]. The proof is a modification of the proof for H^∞. We will use it in the proof of the corona theorem for $H^\infty \cap C_B$.

<u>Theorem 4.1</u>. The closed unit ball of $H^\infty \cap C_B$ is the norm-closed convex hull of the Blaschke products in $H^\infty \cap C_B$.

<u>Proof</u>. Let $h \in H^\infty \cap C_B$ with $\|h\|_\infty \le 1$. For each $\epsilon > 0$, there exist Blaschke products $b_0, b_1, \cdots, b_n$ in C_B, $\lambda_i \in \mathbb{C}$, such that $\|h - \bar{b}_0 \Sigma \lambda_i b_i\| < \epsilon$ and $\|\Sigma \lambda_i b_i\| < 1$. Let $g = \Sigma \lambda_i b_i$. Then $d(g\bar{b}_0, H^\infty) \le \|g\bar{b}_0 - h\| < \epsilon$. By Corollary 2.2 there is a unimodular function $v \in C_B$ with $\epsilon v b_0 = g - b_0 k$ for some $k \in H^\infty \cap C_B$.

We let $N_B = \{f \in H^\infty \cap C_B\colon \bar{f}u \in H^\infty \cap C_B$ for some inner function $u \in H^\infty \cap C_B\}$. Bernard's idea (see [21]) shows that N_B is an algebra containing all inner functions in $H^\infty \cap C_B$ and that the unit ball of the closure of N_B is the closed convex hull of the inner functions in $H^\infty \cap C_B$. We see that $vb_0 \in N_B$, $g \in N_B$ and hence $b_0 k \in N_B$. So $\overline{b_0 k}\cdot u \in H^\infty \cap C_B$ for some inner function $u \in H^\infty \cap C_B$. Thus $\bar{k}u = \overline{b_0 k}u \cdot b_0 \in H^\infty \cap C_B$ and we conclude $k \in N_B$. Finally,

$$\|h - k\| = \|b_0 h - b_0 k\| \le \|b_0 h - g\| + \|g - b_0 k\| < 2\epsilon$$

showing that h is in the unit ball of the closure of N_B. We remark that if u is an inner function in $H^\infty \cap C_B$, then by Frostman's theorem [13], u can be approximated by Blaschke products of the form $(u - \lambda)/(1 - \bar{\lambda}u)$ which are clearly in $H^\infty \cap C_B$.

The proof of the next lemma is exactly the same as in the special case $B = L^\infty$ given in [11].

Lemma 4.2. If $\varphi \in \mathfrak{M}(H^\infty \cap C_B)$ then the following are equivalent:

a) $|\varphi(b)| = 1$ for all Blaschke products $b \in C_B$

b) φ is in the Shilov boundary of $H^\infty \cap C_B$

c) $\varphi \in \mathfrak{M}(C_B)$.

By Theorem 9.5.3 in [20] any $\varphi \in \mathfrak{M}(H^\infty \cap C_B)$ satisfying the above conditions has an extension to an element of $\mathfrak{M}(H^\infty)$ since C_B is a C^*-algebra.

Theorem 4.3. The canonical map $\pi: \mathfrak{M}(H^\infty) \to \mathfrak{M}(H^\infty \cap C_B)$ is onto. In other words, Δ is dense in the maximal ideal space of $H^\infty \cap C_B$.

Proof. We first make a reduction of the problem, the idea of which comes from [2]. If $\varphi \in \mathfrak{M}(H^\infty \cap C_B)$ and $\varphi \notin \pi(\mathfrak{M}(H^\infty))$ then by Lemma 4.2, $|\varphi(b)| < 1$ for some Blaschke product b in C_B. If $\psi \in \mathfrak{M}(H^\infty \cap C_B)$ and $\psi \neq \varphi$, we claim there is an inner function u_ψ in $H^\infty \cap C_B$ such that $\varphi(u_\psi) = 0$ and $\psi(u_\psi) \neq 0$. Indeed, by Theorem 4.1 we can find a Blaschke product $b_\psi \in H^\infty \cap C_B$ such that $\varphi(b_\psi) \neq \psi(b_\psi)$. Then either $u_\psi = (b - \varphi(b))/(1 - \overline{\varphi(b)}b)$ or

$$u_\psi = \frac{(b+\epsilon)/(1+\epsilon b)b_\psi - \varphi((b+\epsilon)/(1+\epsilon b)b_\psi)}{1 - \overline{\varphi((b+\epsilon)/(1+\epsilon b)b_\psi)}\cdot(b+\epsilon)/(1+\epsilon b)b_\psi}$$

works, for some small $\epsilon > 0$. Since $\pi(\mathfrak{M}(H^\infty))$ is compact, there exist inner functions $u_1, u_2, \cdots, u_n$ in C_B such that $\varphi(u_i) = 0$, $1 \le i \le n$, and $\Sigma|u_i| \ge \delta$, $\delta > 0$, on $\pi(\mathfrak{M}(H^\infty))$. Since $z \in H^\infty \cap C_B$, we have $\Delta \subset \pi(\mathfrak{M}(H^\infty))$, so that $\Sigma|u_i| \ge \delta > 0$ on Δ.

We now obtain a contradiction as follows. By the corona theorem for H^∞[4], there exist $g_1, \cdots, g_n$ in H^∞ such that $\Sigma u_i g_i = 1$. Let $u = \Pi u_i$. By Corollary 2.2, $\bar{u}u_i g_i = h_i + v_i$, $1 \le i \le n$, where $h_i \in H^\infty$ and $v_i \in C_B$ with $\|v_i\| = 2\|g_i\|$. So $1 = \Sigma u_i g_i = u\Sigma h_i + \Sigma u_i(u\bar{u}_i v_i)$. Since u and $u\bar{u}_i v_i$ are in $H^\infty \cap C_B$, we have $\Sigma h_i \in H^\infty \cap C_B$. Finally we see that

$$\Sigma u_i p_i = 1$$

where $p_1 = u\bar{u}_1(\Sigma h_i + v_1) \in H^\infty \cap C_B$ and $p_i = u\bar{u}_i v_i \in H^\infty \cap C_B$. This gives the contradiction $1 = \varphi(1) = \Sigma\varphi(u_i)\varphi(p_i) = 0$. This proves the theorem.

5. Further Results. As in Marshall [23] for the case $B = L^\infty$, Theorems 4.1 and 3.2 yield the following corollary.

Corollary 5.1. Let D be a closed algebra between $H^\infty \cap C_B$ and C_B. The unit ball of D is the closed convex hull of $\{b_1\bar{b}_2$: b_1 and b_2 are Blaschke products in $H^\infty \cap C_B$ and b_2 is an interpolating Blaschke product with $\bar{b}_2 \in D\}$.

Proof. It is easy to see from Theorems 4.1 and 3.2 that the corollary is true if b_2 is a finite product of interpolating Blaschke products. By moving the zeros of b_2 slightly, we get an interpolating Blaschke product close to b_2. This new Blaschke product will be invertible wherever b_2 is invertible.

The main results of this paper yield a structure theorem for algebras between $H^\infty \cap C_B$ and C_B which is analogous to (and contains) Theorem 2.4. If D is a closed algebra between $H^\infty \cap C_B$ and C_B, let $C_{B,D}$ be the closed C^*-algebra generated by the invertible Blaschke products in D.

Theorem 5.2. $D = H^\infty \cap C_B + C_{B,D}$.

We outline the proof which follows the lines of the proof of Theorem 2.4. First we remark that $B_1 = [H^\infty \cap C_B, \bar{z}]$ is the smallest algebra between $H^\infty \cap C_B$ and C_B. Indeed, any subalgebra D of C_B that contains $H^\infty \cap C_B$ properly, must contain the complex conjugate of a Blaschke product by Theorem 3.2. By the same argument as given on page 6, $\bar{z} \in D$.

Lemma 5.3. If v is a unimodular function in C_B with $d(v, H^\infty \cap C_B) = 1$ and $d(v, B_1) < 1$, then $\bar{v} \in [H^\infty \cap C_B, v]$.

Proof. The proof is the same as the proof of Theorem 2.1 with H^∞ replaced by $H^\infty \cap C_B$ and $[H^\infty, \bar{z}]$ replaced by B_1. The function $h \in H^\infty \cap C_B$ in the proof is invertible in H^∞ as before. By Theorem 4.3, h is also invertible in $H^\infty \cap C_B$.

As mentioned before, the analogue of Nevanlinna's theorem is false, but we can avoid it.

Lemma 5.4. If $g \in H^\infty \cap C_B$ and if b is an inner function in $H^\infty \cap C_{B,D}$ with $d(g\bar{b}, H^\infty \cap C_B) < 1$, then there exists a unimodular function $u \in g\bar{b} + H^\infty \cap C_B$ with $u \in C_{B,D}$. Furthermore, if $g \in H^\infty \cap C_{B,D}$ then $u \in g\bar{b} + H^\infty \cap C_{B,D}$.

Proof. As in Corollary 2.2, there exists a unimodular function $u \in C_B$ with $u \in g\bar{b} + H^\infty \cap C_B$ and $d(u, zH^\infty) = 1$. Now $d(\bar{z}u, H^\infty \cap C_B) = d(\bar{z}u, H^\infty) = 1$ and $d(\bar{z}u, B_1) \leq d(u, H^\infty \cap C_B) = d(g\bar{b}, H^\infty \cap C_B) < 1$. So by Lemma 5.3, $z\bar{u} \in [H^\infty \cap C_B, \bar{z}u]$, and we see that $\bar{u} \in [H^\infty \cap C_B, u]$. But $bu = g + bh$ for some $h \in H^\infty \cap C_B$. Thus $\overline{bu} \in [H^\infty \cap C_B, g\bar{b}, \bar{b}] \subset D$ and so $bu \in H^\infty \cap C_{B,D}$. Since $b \in C_{B,D}$ we must have $u \in C_{B,D}$. The last statement of the corollary is clear.

The analogue of Lemma 2.5 and theconclusion of the proof of Theorem 5.2 is exactly the same as in section 2, with the appropriate modifications of the algebras, using the above lemmas and Theorem 3.2.

REFERENCES

1. A. Bernard, J. B. Garnett and D. E. Marshall, Algebras generated by inner functions, to appear in Jour. Funct. Anal.
2. A. Bernard, J. B. Garnett and D. E. Marshall, The algebra generated by inner functions, unpublished manuscript.
3. L. Carleson and S. Jacobs, Best uniform approximation by analytic functions, Ark. Math., 10 (1972), 219-229.
4. L. Carleson, The corona theorem, Proceedings of the 15th Scandinavian Congress, Oslo, 1968. Lecture Notes in Mathematics, 118, Springer-Verlag.
5. L. Carleson, Interpolations by bounded analytic functions and the corona problem, Ann. Math., 76 (1962), 547-559.
6. S. Y. Chang, A characterization of Douglas subalgebras, Acta Math., 137 (1976), 81-89.
7. S. Y. Chang, Structure of subalgebras between L^∞ and H^∞, to appear in Trans. Amer. Math. Soc.
8. A. M. Davie, T. W. Gamelin and J. B. Garnett, Distance estimates and pointwise bounded density, Trans. Amer. Math. Soc., 175 (1973), 37-68.
9. D. Dawson, Stable subalgebras of H^∞, preprint, Univ. of Alberta, Edmonton, Alberta, Canada.
10. J. Detraz, Algèbres de fonctions analytiques dans le disque, Ann. Scient. Éc. Norm. Sup. 4e Seine t.3 (1970), 313-352.
11. R. G. Douglas and W. Rudin, Approximation by inner functions, Pacific Jour. Math., 31 (1969), 313-320.
12. S. Fisher, The convex hull of the finite Blaschke products, Bull. Amer. Math. Soc., 74 (1968), 1128-1129.
13. O. Frostman, Potential d'équilibre et capacité des ensembles avec quelques applications à la théorie des fonctions, Meddel, Lunds Univ. Mat. Sem. 3 (1935), 1-118.
14. C. Fefferman and E. M. Stein, H^p spaces of several variables, Acta Math., 129 (1972), 137-193.
15. T. W. Gamelin and J. Garnett, Uniform approximation by bounded analytic functions, Revista de la Union Matématica Argentina, 25 (1970).
16. J. B. Garnett, Interpolating sequences for bounded harmonic functions, Ind. Univ. Math. Jour., 21 (1971), 187-192.
17. N. Jewell, thesis, Univ. of Edinburgh, Scotland.
18. K. Hoffman, Banach Spaces of Analytic Functions. Prentice Hall, Englewood Cliffs, NJ, 1962.
19. K. Hoffman and I. M. Singer, Maximal subalgebras of continuous functions, Acta Math., 103 (1960), 217-241.
20. R. Larsen, Introduction to Banach Algebras, Series on Pure and Applied Mathematics, 24 (1973).
21. D. Marshall, Blaschke products generate H^∞, Bull. Amer. Math. Soc., 82 (1976), 494-496.
22. D. Marshall, Subalgebras of L^∞ containing H^∞, Acta Math., 137 (1976), 91-98.
23. D. Marshall, thesis, Univ. of Calif. Los Angeles, CA, 1976.
24. R. Nevanlinna, Über beschränkte analytische Funktionen, Ann. Acad. Sci. Fenn., Ser. A, 32, 7 (1929).
25. D. Sarason, Algebras of functions on the unit circle, Bull. Amer. Math. Soc., 79 (1973), 286-299.
26. D. Sarason, Functions of vanishing mean oscillation, Trans. Amer. Math. Soc., 207 (1975), 391-405.
27. D. Sarason, Algebras between L^∞ and H^∞, Spaces of Analytic Functions, Kristiansand, Norway 1975, Springer Lecture Notes 512.
28. J. Wermer, On algebras of continuous functions, Proc. Amer. Math. Soc., 4 (1953), 866-869.
29. S. Ziskind, Interpolating sequences and the Shilov boundary of $H^\infty(\Delta)$, Jour. Funct. Anal., 21 (1976), 380-388.

A THEOREM ON COMPOSITION OPERATORS

Joseph A. Cima
University of North Carolina
Chapel Hill, NC 27514/USA

1. Introduction. We shall use the usual notation of Hardy space theory, see Duren [3]. Hence if ϕ is a function in the unit ball of H^∞ then the composition operator induced on H^2 by ϕ is $C_\phi(f)(z) = f(\phi(z))$. There is a substantial literature on the subject, see for example, [2], [4], and [5] and the references therein. It is the purpose of this paper to obtain a result analogous to Theorem 1 of [2] for the Dirichlet space. Our proof is patterned after that in [2] but has some essentially different mathematical ingredients. In particular, we appealed to two facts which are valid in H^2. The first trivial one is that multiplication of H^2 by an H^∞ function yields a closed subspace of H^2. The second fact we needed in our proof was the Theorem of Szego. We do not have these tools available in the case of the Dirichlet space and indeed the latter fact raises the question: if f is a positive function in $\bigcap_{1<p} L^p$, when can one find a g with a finite Dirichlet integral such that $|g(e^{i\theta})| = f(e^{i\theta})$, a.e.?

The Dirichlet space $\mathfrak{D}$ consists of the functions f analytic on the unit disk Δ, with finite Dirichlet integral

$$\|f\|_{\mathfrak{D}}^2 = \int_\Delta \int |f'(z)|^2 \, dxdy.$$

We always assume $f(0) = 0$ in $\mathfrak{D}$ and since $\mathfrak{D} \subset H^2$ the usual factorization properties hold for functions in $\mathfrak{D}$. We will occasionally use $\| \ \|_{\mathfrak{D}}$ for functions with finite Dirichlet integral which do not vanish at the origin.

2. The Characterization of Fredholm Composition Operators on $\mathfrak{D}$. If $f(z) = B(z)S(z)F(z)$ is a function in $\mathfrak{D}$ it is not always possible to conclude that any of the individual pieces of this composition are in $\mathfrak{D}$. For example $f(z) = z \exp(\frac{z+1}{z-1}) \log(1 - z)$ is in $\mathfrak{D}$. The result we need is contained in a proof of a theorem of L. Carleson [1]. In particular, Carleson has characterized the norm of an f in $\mathfrak{D}$ in terms of integrals of the components occuring in its factorization.

Theorem A. (L. Carleson). Let f(z) be analytic in $|z| < 1$ with a finite Dirichlet integral $\mathfrak{D}(f)$

$$\mathfrak{D}(f) = \int\int_{|z|<1} |f'(z)|^2 \, dxdy.$$

f(z) can be factorized,

$$f(z) = ax^m \prod_1^\infty \frac{\bar{z}_\nu}{|z_\nu|} \left(\frac{z_\nu - z}{1-\bar{z}_\nu z}\right) \cdot \exp\left\{-\int_{-\pi}^{\pi} \frac{e^{i\phi}+z}{e^{i\phi}-z} ds(\phi)\right\}$$

$$\times \exp\left\{\frac{1}{2\pi}\int_{-\pi}^{\pi} \frac{e^{i\phi}+z}{e^{i\phi}-z} \log|f(e^{i\phi})|\,d\phi\right\} .$$

Here s is singular and nondecreasing. We denote the boundary values $|f(e^{i\phi})|$ also $e^{u(\phi)}$. Then

$$\mathfrak{D}(f) = a^2\left\{\frac{1}{2}\int_{\pi}^{\pi} \left(M + \sum_1^\infty \frac{1-|z_\nu|^2}{|e^{i\theta}-z_\nu|^2}\right) |f(e^{i\theta})|^2 d\theta + \right.$$

$$+ \frac{1}{4}\int_{-\pi}^{\pi}\int \frac{|f(e^{i\theta})|^2 ds(\phi)}{\sin^2\frac{\theta-\phi}{2}}\,d\theta +$$

$$\left. + \frac{1}{8\pi}\int_0^{\pi} \frac{dt}{\sin^2\frac{t}{2}} \int_{-\pi}^{\pi} (u(x+t) - u(x))(e^{2u(x+t)} - e^{2u(t)})dx\right\}.$$

If X is a Hilbert space and T is a bounded linear operator on X into X then T is a Fredholm operator if the range of T is closed and if the dimension of the kernel and the cokernel of T are finite.

Theorem 1. Let ϕ be in the unit ball of H^∞. A composition operator C_ϕ on $\mathfrak{D}$ is a Fredholm operator if and only if ϕ is a rotation, $\phi(z) = \lambda z$, $|\lambda| = 1$.

Proof. Assume that $\phi = BSF$ and C_ϕ is a Fredholm operator on $\mathfrak{D}$. Equation (4), page 191 on [1], shows that SF has a finite Dirichlet integral and equation (7), page 193 of [1] shows that F has a finite Dirichlet integral. If

$B(z) = \prod_{\nu=1}^{\infty} \left(\frac{\bar{z}_\nu}{|z_\nu|}\right)\left(\frac{z_\nu - z}{1-\bar{z}_\nu z}\right)$, then $B_n(z)$ is the n-th partial product of this product.

It is easily seen that if $\phi_\alpha(z) = (z-\alpha)(1-\bar{\alpha}z)^{-1}$, $|\alpha| < 1$ and if g has a finite Dirichlet integral then

$$\|\Phi_\alpha g\|_{\mathfrak{D}}^2 = \|Z \cdot (g \circ \Phi_{-\alpha})\|_{\mathfrak{D}}^2 = \|g\|_{\mathfrak{D}}^2 + \|g \circ \Phi_{-\alpha}\|_2^2.$$

The sequence B_nF converges pointwise (and uniformly on compacta) to BF. The norms of B_nF are increasing on n and are bounded above by

$$\int_\Delta\int |F'(z)|^2\,dxdy + \frac{1}{2}\int_{-\pi}^{\pi} \sum_{\nu=1}^{\infty} \frac{1-|z_\nu|^2}{|e^{i\theta}-z_\nu|^2} |F(e^{i\theta})|^2\,d\theta .$$

Hence, $BF \in \mathfrak{D}$.

The kernel of C_ϕ is trivial and the linear span of the powers of ϕ yields the range of C_ϕ. In particular, Range $C_\phi \subseteq B\mathfrak{D} \cap \mathfrak{D}$. The set $B\mathfrak{D} \cap \mathfrak{D}$ is a nontrivial subspace of $\mathfrak{D}$. It is closed by the discussion given above. The reproducing kernels

$$K_{z_\nu}(z) = \log\left(\frac{1}{1 - \bar{z}_\nu z}\right)$$

for $\mathfrak{D}$ are a linearly independent set in $(B\mathfrak{D} \cap \mathfrak{D})^\perp$. Hence, B cannot have an infinite number of distinct zeros.

Now let $\rho(\theta)$ be an increasing function that induces the singular positive measure for the singular factor $S(z)$. We may write $\rho(\theta) = \rho_1(\theta) + \rho_2(\theta)$ as the sum of two increasing functions so that $d\rho_1$ and $d\rho_2$ have nonempty disjoint supports. If $\rho(\theta)$ is a single step function we may take a square root (or n^{th} root) of $S(z)$. Let $S_i(z)$ be the singular function induced by $\rho_i(\theta)$. If $N = S\mathfrak{D} \cap \mathfrak{D}$ and $N_1 = S_1\mathfrak{D} \cap \mathfrak{D}$ then N and N_1 are proper closed subspaces of $\mathfrak{D}$. Further N is a proper closed subspace of N_1. Hence

$$\dim N^\perp > \dim N_1^\perp \geq 1.$$

We see that if $S(z) \not\equiv \lambda$, then we must have codim $C_\phi \mathfrak{D}$ infinite.

We are reduced to the case

$$\phi(z) = \prod_{\nu=1}^{k} \left(\frac{\bar{z}_\nu}{|z_\nu|}\right)\left(\frac{z - z_\nu}{1 - \bar{z}_\nu z}\right)F(z) = B(z)F(z)$$

where k is a positive integer. But the observation $[\phi^n, \phi^{n+1}, \phi^{n+2}, \cdots] \subseteq B^n\mathfrak{D} \cap \mathfrak{D} \equiv M$ shows that since codim $M \geq kn$, the codim range $C_\phi \geq (k-1)n$. This yields $k = 1$, and $\phi(z) = zF(z)$. We shall show that $F(z)$ is a constant. To this end assume there is a compact set $K(\subseteq \Gamma)$ of positive measure, and $|F(e^{i\theta})| \leq \rho < 1$ for $e^{i\theta} \in K$. Choose $p_1 \in K$ and an open interval V_1 containing p_1 such that

$$0 < m(K \cap V_1) < \frac{a}{2},$$

where $mK = a$. Define $K_1 = K\ (K \cap V_1)$. K_1 is compact with measure larger than $a/2$. Choose $p_2 \in K_1$ and an open interval V_2 containing p_2 such that

$$0 < m(V_2 \cap K_1) < \frac{a}{2^2}.$$

Now define $K_2 = K_1 \backslash (K_1 \cap V_2)$ and proceed inductively. Let Ω be the simply connected domain of finite area pictured in Figure 1.

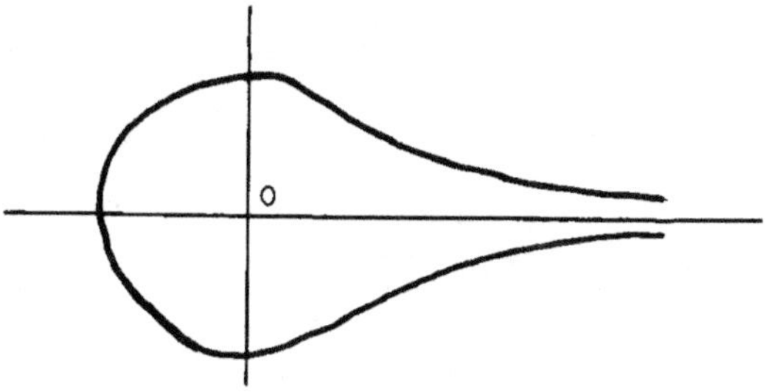

FIGURE 1

We have the existence of a holomorphic, univalent function ψ_j mapping Δ onto Ω, $\psi_j(0) = 0$ and $\psi_j(p_j) = \infty$. Of course, ψ_j is in $\mathfrak{D}$ but not $C_\phi\mathfrak{D}$. Let P be the orthogonal projection of $\mathfrak{D}$ onto the orthogonal complement of the range of C_ϕ. One can show that $P(\psi_j) = g_j$ is a linearly independent set in $(C_\phi\mathfrak{D})^\perp$. This is again a contradiction and thus $\phi(z) = \lambda z$, $|\lambda| = 1$.

The above result suggests the following problem. Characterize those ϕ in the unit ball of H^∞ for which C_ϕ has dense range. It is easily seen from the proof above that $\phi(z)$ must have the factorization $zF(z)$. Furthermore, F is an invertible function in H^∞. Also if C_ϕ has dense range then also for $\psi(z) = r\phi(z)$, $0 < r < 1$, we see that C_ψ has dense range. We offer no conjecture on this question.

REFERENCES

1. L. Carleson, A representation formula for the Dirichlet integral, Math. Zeit., 73 (1960), 190-196.
2. J. A. Cima, J. Thomson, and W. Wogen, On some properties of composition operators, Indiana Univ. Math. Jour. 24 (1974), 200-215.
3. Peter L. Duren, Theory of H^p spaces. New York/London: Academic Press, 1970.
4. H. J. Schwartz, Composition operators on H^p, Thesis, University of Toledo, 1969.
5. J. H. Shapiro and P. D. Taylor, Compact, nuclear, and Hilbert-Schmidt composition operators on H^2, Indiana Univ. Math. Jour., 23 (1973), 471-496.

THE DISTANCE OF SYMMETRIC SPACES FROM $\ell_p^{(n)}$

W. Davis
Ohio State University
Columbus, OH 43210/USA

Let E and F be Banach spaces. The Banach-Mazur distance from E to F is defined by $d(E, F) = \inf \|T\|\|T^{-1}\|$ where the infimum is taken over all isomorphisms T of E onto F. Geometrically, $d(E, F) \leq C$ if there is a linear map T of E onto F with $B(F) \subset T(B(E)) \subset CB(F)$, where B(E) denotes the unit ball of E. Relatively little is known about distances between finite dimensional spaces. F. John [4] proved that, if dim E = n then $d(E, \ell_2^n) \leq \sqrt{n}$. It follows that $D(E, F) \leq n$ whenever dim E = dim F = n. There is no sequence of pairs of Banach spaces (E_n, F_n) (with dim E_n = dim F_n = n) for which it is known that $\sqrt{n} = o(d(E_n, F_n))$.

Here we restrict our attention to symmetric spaces. A space E is symmetric if E has a basis $(e_i)_{i=1}^n$ such that

$$\|\Sigma\, a_i e_i\| = \|\Sigma\, \epsilon_i a_i e_{\pi(i)}\| \text{ for all } (a_i)_{i=1}^n,$$

all $(\epsilon_i = \pm 1)_{i=1}^n$ and all permutations π of $\{1, \cdots, n\}$. Gurarii, Kadec and Macaev in [2] and [3] proved the following results in this context:

1. $$d(\ell_p^n, \ell_q^n) \leq \begin{cases} C_{p,q}\, n^{\left|\frac{1}{p} - \frac{1}{q}\right|} & \text{if } 1 \leq p, q \leq 2 \text{ or } 2 \leq p, q \leq \infty \\ C_{p,q} \max\{n^{\left|\frac{1}{p} - \frac{1}{2}\right|}, n^{\left|\frac{1}{q} - \frac{1}{2}\right|}\} & \text{if } 1 \leq p \leq 2 \leq q \leq \infty, \end{cases}$$

   where $C_{p,q}$ is a constant not depending on the dimension of the spaces.
2. If E is symmetric, then $d(E, \ell_1^n)\, d(E, \ell_\infty^n) \leq n$, and
3. If E is symmetric, then $d(E, \ell_1^n) \leq c\, n^{3/4}$ (by duality also $d(E, \ell_\infty^n) \leq C\, n^{3/4}$).

In this note we extend the last result to: If E is symmetric, then $d(E, \ell_p^n) \leq D\sqrt{n \log n}$, for every p, $1 \leq p \leq \infty$.

For the arguments which follow, we shall restrict our attention to $n = 2^k$ and the real field of scalars. The extension to general n is formal and will be done at the end. Also, the solution in the complex case is easier (with better constants) in most situations since there are, for every n, complex orthogonal matrices (a_{ij}) with $|a_{ij}| = 1$ for $1 \leq i, j \leq n$.

First we give a proof of the first cited result of Gurarii, Kadec and Macaev. In the case $1 \leq p < q \leq 2$, the identity map of ℓ_p^n into ℓ_q^n yields the desired upper

---

This work supported by NSF grant number MCS-74-07509A02.

bound: First $B(\ell_p^n) \subset B(\ell_q^n)$. Second, let $(a_1, \cdots, a_n) \in B(\ell_q^n)$. By Hölder's inequality $(\Sigma |a_i|^p)^{1/p} \le (\Sigma |a_i|^q)^{1/q} n^{\frac{1}{r}}$ where $\frac{1}{q} + \frac{1}{r} = \frac{1}{p}$, that is, $\frac{1}{r} = \frac{1}{p} - \frac{1}{q}$. This shows that $B(\ell_q^n) \subset n^{\frac{1}{p}-\frac{1}{q}} B(\ell_p^n)$, so that $d(\ell_p^n, \ell_q^n) \le n^{\frac{1}{p}-\frac{1}{q}}$. The case $2 \le q \le p \le \infty$, the mapping to be used is the Walsh matrix ($W_1 = (1)$, $W_{k+1} = \begin{pmatrix} W_k & W_k \\ W_k & -W_k \end{pmatrix}$) suitably normalized. For the sake of definiteness, assume that $q' \ge p$ where $\frac{1}{q} + \frac{1}{q'} = 1$, and let $T = n^{-\frac{1}{q}} W_k: \ell_p^{(n)} \to \ell_q^{(n)}$. We claim 1. $B(\ell_q^{(n)}) \subset n^{\frac{1}{p}-\frac{1}{2}} T(B(\ell_p^{(n)}))$, and 2. $\|T\| \le 1$ so that $T(B(\ell_p^{(n)})) \subset B(\ell_q^{(n)})$. To see (1), notice that $B(\ell_q^{(n)}) \subset n^{\frac{1}{2}-\frac{1}{q}} B(\ell_2^{(n)})$, and that $T(B(\ell_2^{(n)})) \supset n^{\frac{1}{2}-\frac{1}{q}} B(\ell_2^{(n)})$ since $n^{-\frac{1}{2}} W_k$ is an orthogonal matrix and $n^{\frac{1}{2}-\frac{1}{q}} \le n^{\frac{1}{p}-\frac{1}{2}}$. On the other hand, $n^{\frac{1}{p}-\frac{1}{2}} B(\ell_p^{(n)}) \supset B(\ell_2^{(n)})$, so that $T(n^{\frac{1}{p}-\frac{1}{2}} B(\ell_p^{(n)})) \supset T(B(\ell_2^{(n)})) \supset B(\ell_q^{(n)})$, as desired. For (2), we substitute an application of Clarkson's inequality for the use of the Riesz interpolation theorem in [2]: Proceed by induction on k. Clearly, for $k = 0$, T_0: $\ell_p^{(1)} \to \ell_q^{(1)}$ has norm 1. Suppose we know that $T_k (\equiv T)$ is norm 1 from $\ell_p^{(n)}$ to $\ell_q^{(n)}$. Then, by definition, T_{k+1} has a matrix representation of the form $2^{-\frac{1}{q}} \begin{pmatrix} T_k & T_k \\ T_k & -T_k \end{pmatrix}$. Let $u \in \ell_p^{(2n)}$ and write $u = (x, y)$ where $x = (u_1, \cdots, u_n)$, $y = (u_{n+1}, \cdots, u_{2n})$. Then $T_{k+1}(u) = 2^{-\frac{1}{q}}(T_k(x + y), T_k(x - y))$. Clarkson's second inequality [1] implies that the map S: $\ell_{q'}^{(2)} \to \ell_q^{(2)}$ defined by $S(\alpha, \beta) = (\alpha + \beta, \alpha - \beta)$ has norm $2^{\frac{1}{q}}$. It then follows that S, as a map from $\ell_p^{(2)}$ to $\ell_q^{(2)}$ also has norm $2^{\frac{1}{q}}$ when $p \le p'$ (as we have here). This forces $\|T_{k+1}\| \le 1$, as desired.

Now we proceed to the proof of the main results of this paper. We need the following lemma:

<u>Lemma 1</u>: Let E be an n-dimensional symmetric space, $n = m \cdot \ell$, $m = 2^j$ and let $x = (x_i)_{i=1}^n$ with $x_i = x_1 \ne 0$ for $1 \le i \le m$ and $x_i = 0$ for $i > m$. Then (with the usual inner product on E considered as R^n), there is an orthogonal basis for E in the orbit of x under the symmetry group of E.

<u>Proof</u>: Let $X_j = x_1 W_j$, which has orthogonal rows. Then the matrix

$$X = \begin{pmatrix} X_j & \cdots & 0 \\ & & \\ 0 & \cdots & X_j \end{pmatrix}$$

has orthogonal rows, and each row is in the orbit of x as desired.

<u>Theorem 1</u>: If E is an n-dimensional symmetric space ($n = 2^k$), then $d(E, \ell_1^{(n)}) \le K\sqrt{n \log n}$.

Proof: Assume, with no loss of generality, that $E = \langle R^n, \|\cdot\|\rangle$, that $B(E) \subset B(\ell_2^{(n)})$ with the usual inner product, and that there exists $x \in E$, $\|x\|_E = \|x\|_2 = 1$. By symmetry, we may as well assume that $x = (x_1, x_2, \cdots, x_n)$ with $x_1 \geq x_2 \geq \cdots \geq x_n \geq 0$. Let $m = 2^j$, and $S(y) = \sum_{i=1}^{n} y_{\rho(i)} e_i$, where ρ is the permutation

$$\rho = (2, \cdots, m, 1)(m + 2, \cdots, 2m, m + 1) \cdots (n - m + 2, \cdots, n, n - m+1).$$

Let $U(y) = \Sigma_1^m y_i e_i - \Sigma_{i=m+1}^n y_i e_i$. S and U are isometries of E, so $\|y^{(m)}\| \leq \|y\|$ where $y^{(m)} = \frac{1}{2m}(I-U)\sum_{i=0}^{m-1} S^i(y)$. Also, $y^{(m)}$ has an orthogonal basis, say $(Y_{m,i})_{i-1}$, in its orbit, by lemma 1. If we let $\|y\| \leq 1$, let (δ_i) denote the unit vector basis of $\ell_1^{(n)}$ and define a map $T: \ell_1^{(n)} \to E$ by $T(\delta_i) = Y_{m,i}$; $i = 1, \cdots, n$, then we have $T(B(\ell_1^{(n)})) \subset \|y^{(m)}\|_2 B(\ell_2^{(n)}) \subset \sqrt{n}\, \|y^{(m)}\|_2 T(B(\ell_1^{(n)}))$ by direct computation. Thus, $B(\ell_2^{(n)}) \subset \|y^{(m)}\|_2^{-1} \sqrt{n}\, T(B(\ell_1^n))$. Also, since $\|y^{(m)}\| \leq 1$, $T(B(\ell_1^{(n)})) \subset B(E)$. Since, $B(E) \subset B(\ell_2^{(n)})$, we have, finally, $T(B(\ell_1^{(n)})) \subset B(E) \subset \|y^{(m)}\|_2^{-1} \sqrt{n}\, T(B(\ell_1^{(n)}))$, so $d(E, \ell_1^{(n)}) \leq \|y^{(m)}\|_2^{-1} \sqrt{n}$. Thus, to get the desired upper bound on this distance, we will examine small values of $\|y^{(m)}\|_2^{-1}$. Let $\|x\|_E = \|x\|_2 = 1$, and $x^{(0)}$, $x^{(1)}$, $x^{(2)}$, $x^{(4)}$, etc., be defined as above. We wish to estimate $M_x = \max_{0 \leq j \leq k} \{\|x^{(m)}\|_2 : m = 2^j\}$. We estimate $\mu = \min\{M_x : \|x\|_2 \geq 1, x_1 \geq x_2 \geq \cdots \geq 0\}$. For any $x (\geq 0)$, $\|x^{(m)}\|_2 = \dfrac{x_1 + \cdots + x_m}{\sqrt{m}}$. It is routine that μ is attained for x satisfying $\|x^{(1)}\|_2 = \|x^{(m)}\|_2$ for all m $(= 2^j)$, so that

(1) $\displaystyle\sum_{i=2^{j-1}+1}^{2^j} x_i = (\sqrt{2} - 1)2^{\frac{j-1}{2}} x_1$. Since $x_1 \geq x_2 \geq \cdots$, it must be that

(2) $\displaystyle x_i \leq 2^{-j} \left(\sum_{i=1}^{2d} x_i\right) = 2^{-\frac{j}{2}} x_1$ for $2^j < i$.

Using just (1) and (2), and $\|x\|_2 \geq 1$, we see that

$$1 \leq \sum_{i=1}^{n} x_i^2 \leq x_1^2 \left(1 + \sum_{j=1}^{k} [(\sqrt{2} - 1)\, 2^j]\, 2^{-j}\right) = x_1^2 (1 + (\sqrt{2} - 1) \log_2 n).$$

Thus, $\mu = x_1 \geq (1 + (\sqrt{2} - 1) \log_2 n)^{-1/2}$. That is, we have shown that, for any $x \in E$, $\|x\| = \|x\|_2 = 1$, there is $m = 2^j$ such that $\|x^{(m)}\|_2^{-1} \leq (1 + (\sqrt{2} - 1)\log_2 n)^{1/2}$, completing the proof.

Theorem 2: If E is an n-dimentional symmetric space $(n = 2^k)$, then $d(E, \ell_p^{(n)}) \leq K\sqrt{n \log n}$.

Proof: Since $d(F, \ell_p^{(n)}) = d(F^*, \ell_{p'}^{(n)})$ if F is an n-dimensional space, we shall assume, without loss of generality that $1 \le p \le 2$. Again assume that $B(E) \subset B(\ell_2^{(n)})$ and that there is $x \in B(E)$ with $\|x\|_2 = 1$. In the proof of theorem 1 we found an orthogonal basis $(x_i)_{i=1}^n$ in E satisfying $\|x_i\|_E \le 1$, $(1 + (\sqrt{2} - 1) \log_2 n)^{-1/2} \le \|x_i\|_2 \le 1$. Let (δ_i) denote the unit vector basis of $\ell_p^{(n)}$ and define $T: \ell_p^{(n)} \to E$ by $T(\delta_i) = x_i$. Then $T(B(\ell_p^n)) \subset \|x_i\|_2 \, B(\ell_2^n)$, so that (by orthogonality) $\|x_i\|_2^{-1} n^{1/p - 1/2} T(B(\ell_p^n)) \supset B(\ell_2^n) \supset B(E)$. On the other hand, $n^{1/p - 1} n^{1/p - 1} B(\ell_p^n) \subset B(\ell_1^n)$ and $T(B(\ell_1^{(n)})) = co(\pm T(\delta_i))_{i=1}^n \subset B(E)$, so that $n^{1/p - 1} T(B(\ell_p^{(n)})) \subset B(E)$. Putting these inclusions together, we obtain the desired result.

Finally, we see that having found a bound on the distance for $n = 2^k$, that bound gives the correct order of magnitude for all dimensions. Notice also that obtaining this bound for $\ell_1^{(n)}$ is sufficient in the proof of theorem 2 for all p.

<u>Proposition</u>: Suppose $\|\cdot\|$ defines a symmetric norm on R^j for all j, and that there is a concave $f(\cdot)$ satisfying a Δ_2 condition* so that for $n = 2^k$ we have $d(\langle R^n, \|\cdot\|\rangle, \ell_1^n) \le f(n)$. Then there exists λ such that for all m we have $d(\langle R^m, \|\cdot\|\rangle, \ell_1^m) \le \lambda \, f(m)$.

<u>Proof</u>: By induction, assume that we know that $d(\langle R^j, \|\cdot\|\rangle, \ell_1^j) \le \lambda \, f(j)$ for $1 \le j \le 2^k$, and let $m = 2^k + j$ for some such j. Then $E = \langle R^m, \|\cdot\|\rangle = \langle R^{2^k}, \|\cdot\|\rangle + \langle R^j, \|\cdot\|\rangle$ and there are maps T_i, $\|T_i^{-1}\| \le 1$, $T_1: \langle R^{2^k}, \|\cdot\|\rangle \to \ell_1^{2^k}$, $T_2: \langle R^j, \|\cdot\|\rangle \to \ell_1^j$ such that $\|T_1\| \le f(2^k)$ and $\|T_2\| \le \lambda f(j)$. Letting $T = T_1 \oplus T_2$, and $u \in E$, we see that

$$\|Tu\| = \|(T_1(u_1), T_2(u_2))\| = \|T_1 u_1\| + \|T_2 u_2\| \le f(2^k)\|u_1\| + \lambda f(j)\|u_2\| \le (f(2^k) + \lambda f(j)) \, \|u\|.$$

By the Δ_2 condition there exists λ such that $f(2^k) + \lambda f(j) \le \lambda f(2^k + j)$, completing the proof.

Clearly, the author does not know whether or not the bound $\sqrt{n \log n}$ is best possible. It is easy to see that if E is (a Lorentz sequence space) $d(1, \alpha)$, then $d(\ell_1^{(n)}, E) \le O(\sqrt{n})$. However, it does not seem to be known that $d(\ell_\infty^{(n)}, E) \le O(\sqrt{n})$ in this case.

REFERENCES

1. J. A. Clarkson, Uniformly convex spaces, Trans. AMS, 40 (1936), 396-414.
2. V. Gurarii, M. Kadec, and V. Macaev, On the distance between finite dimensional ℓ_p spaces, Mat. Sb., 70 (1966), 481-489.
3. ______, On constants associated with asymmetry of Minkowski spaces, Mat. Sb., 71 (1966), 24-29.
4. F. John. Extremum problems with inequalities as subsidiary conditions, Courant Anniversary Volume, Interscience, New York (1948), 187-204.

*) We use a Δ_2 condition of the form $f(x) \le \delta f(2x)$ with $\delta < 1$ for all x.

WEAKLY CONVERGENT SEQUENCES OF BANACH SPACE VALUED RANDOM VARIABLES

W. J. Davis[(1)] and W. B. Johnson[(2)]
The Ohio State University
Columbus, OH 43210/USA

Abstract: Examples are constructed of sequences $\{X_n\}$ of E valued random variables such that (a) for each $x^* \in E^*$, $x^*(X_n) \to 0$ a.s. and yet (b) X_n does not go to zero weakly a.s.

Chacon and Sucheston [1] recently proved that an L_∞^E bounded sequence $\{X_n\}$ of random variables such that $E(X_\tau)$ converges as $\tau \to \infty$ in the bounded stopping times, converges weakly almost surely if both E and E^* have the Radon-Nikodym property. (Sequences as above are called amarts.) The restriction on E is clearly necessary, since if E fails the Radon-Nikodym property there are E valued martingales which fail to converge weakly a.s. [2]. Here we show that the restriction on E^* is also necessary. We construct examples in ℓ_1, $C(\Delta)$ and the James tree space, [3], of sequences $\{X_n\}$ as in the abstract, and we believe that any E whose dual fails the Radon-Nikodym property possesses such a sequence.

Example 1: Let $E = \ell_1$, so E has the Radon-Nikodym property and E^* fails to have it. Let $\{e_i\}$ denote the natural unit vector basis in ℓ_1 and let $\{\varepsilon_n\}$ be any Bernoulli sequence of random variables (on, say, (Ω, P)), i.e. $\{\varepsilon_n\}$ are independent and $P[\varepsilon_n = 1] = P[\varepsilon_n = -1] = \frac{1}{2}$ for all n. Define

$X_n = 2^{-n} \sum_{j=0}^{2^n-1} e_{2^n+j}\, \varepsilon_{2^n+j}$. For each $\omega \in \Omega$, $\|X_n(\omega)\| = 1$. Further, for each ω, $\{X_n(\omega)\}$ is not norm convergent in ℓ_1, hence not weakly convergent there, so there is $x_\omega^* \in \ell_1^*$ such that $x_\omega^*(X_n(\omega))$ fails to converge. That is, for any subset A of Ω with $P(A) > 0$, there is x^* in E^* and $\omega \in A$ so that $x^*(X_n(\omega))$ does not converge. On the other hand, for any x^* in E^*, $\{x^* \circ X_n\}$ is a scalar, independent, L^∞ bounded sequence which converges to 0 a.s., so we have (a) and (b) of the abstract. It is easy to check directly that $\{X_n\}$ is an amart: Let τ be a large stopping time, e.g., $\tau = \sum_{i \geq N} i\, 1_{A_i}$, and let $x^* \in E^*$, $\|x^*\| \leq 1$. Then, by Chebyshev's inequality, $P[|x^*X_n| \geq \lambda] \leq 2^{-n}\lambda^{-2}$ so that $\mathfrak{S}(|x^*X_i|\, 1_{A_i}) \leq \varepsilon P(A_i) + 2^{-i}\varepsilon^{-1}$ for all $\varepsilon > 0$. Thus, $|\mathfrak{S}(x^*X_\tau)| \leq \varepsilon + \frac{1}{\varepsilon} 2^{-N+1}$, proving the assertion.

The second example is the prototype for further constructions. It is carried out in $C(\Delta)$ (even though that space fails the Radon-Nikodym property and contains

(1) Supported in part by NSF-MCS-74-07509.
(2) Supported in part by NSF-MCS-72-04634.

ℓ_1) without benefit of the ℓ_1 inside the space.

Example 2: Let $E = C(\Delta)$. Let Δ denote the Cantor set, $\{A_{n,k}\}_{n=0}^{\infty}, {}_{i=0}^{2^n-1}$ is natural dyadic decomposition, and P the natural probability measure on $(\Delta, A_{n,i})$, i.e., $P(A_{n,i}) = 2^{-n}$ for all n, i. Define $X_n: \Delta \to C(\Delta)$ by $(X_n(\delta))(t) = \sum_{i=0}^{2^n-1} 1_{A_{n,i}}(\delta)\, 1_{A_{n,i}}(t)$. For each $\delta' \in \Delta$, $(X_n(\cdot))(\delta') = 1_{A_{n,i(\delta')}}(\cdot)$ where $i(\delta')$ is the unique index for which $A_{n,j} \supset \{\delta'\}$. It is clear, therefore, that $(X_n(\delta))(\delta') \to 0$ as $n \to \infty$ if $\delta \neq \delta'$, and is, therefore, routine that $\{x^* \cdot X_n\}$ goes to zero almost surely for each x^* in E^*. On the other hand, for each $\delta \in \Delta$, there is a functional x_δ^* (namely, evaluation at δ) such that $x_\delta^*(X_n(\delta)) \not\to 0$. Thus, $\{X_n\}$ fails to go to zero weakly almost surely. It is once again easy to verify that $\{X_n\}$ is an amart.

It seems possible that this example will give a method for constructing an example satisfying (a) and (b) in any space whose dual fails the Radon-Nikodym property. Any such space contains a separable subspace E with nonseparable dual. Stegall [4] has proved, then, that there is a weak* homeomorph Δ of the Cantor set in the ball of E^*, and there are elements $\{x_{n,i}\}_{n=0,i=0}^{\infty \;\; 2^n-1}$ in the $1+\epsilon$ ball of E such that if $T: E \to C(\Delta)$ is defined by $(Tx)(\delta) = \delta(x)$, then $\sum_{n=0}^{\infty}\sum_{i=0}^{2^n-1} \|Tx_{n,i} - 1_{A_{n,i}}\| < \epsilon$. The candidate for the example is, obviously, $X_n: \Delta \to E$ defined by $X_n = \sum_{i=0}^{2^n-1} x_{n,i} 1_{A_{n,i}}$. The difficulty here is that consideration of the evaluation functionals $\delta \in \Delta$ is no longer sufficient to guarantee that $x^* X_n \to 0$ a.s. for each $x^* \in E^*$. If, however, $\Delta \subset B_{E^*}$ can be chosen so that span $A + \eta$ is dense in where $\eta = \{x^* \in E^* | x^*(x_{n,i}) \to 0$ as $(n, i) \to \infty\}$, then X_n satisfies (a) and (b).

Example 3: In [3], R. C. James constructed an example of a space E which is separable, does not contain ℓ_1 and which has nonseparable dual. It has a doubly indexed basis $\{x_{n,i}\}_{n=0,i=0}^{\infty \;\; 2^n-1}$ as in the Stegall construction, and its Δ (as in Stegall's construction) does have the property that span $\Delta + \eta$ is dense in E^*. Thus X_n as defined satisfies (a) and (b). In fact, X_n is an amart. This is readily checked using the fact that for each n, $\{x_{n,i}\}_{i=0}^{2^n-1}$ is equivalent to the natural basis of $\ell_2^{2^n}$. Therefore, the James tree space is a separable space with the Radon-Nikodym property, with dual failing RNP, not containing ℓ_1 and possessing an L_∞ bounded amart satisfying (a) and (b).

REFERENCES

1. R. V. Chacon and L. Sucheston, On convergence of vector valued asymptotic martingales, Z. Warsch. 33 (1975), 55-59.
2. S. Chatterji, Martingale convergence and the Radon-Nikodym theorem, Math. Scand. 22 (1968), 21-41.
3. R. C. James, A separable somewhat reflexive space with nonseparable dual, B.A. M.S., 80 (1974), 738-743.
4. C. Stegall, The Radon-Nikodym property in conjugate Banach spaces," T.A.M.S., 20 (1975), 213-223.

TWO REMARKS ON INTERPOLATION BY BOUNDED ANALYTIC FUNCTIONS

John B. Garnett*
University of California
Los Angeles, CA 90024/USA

1. Let H^∞ be the ring of bounded analytic functions on the unit disc Δ. Using Fatou's theorem on radial limits, we identify H^∞ with a closed subalgebra of $L^\infty(d\theta/2\pi)$, the bounded Lebesgue measurable functions on $\partial\Delta$. Thus an inner function is a function $f \in H^\infty$ such that $|f(\theta)| = 1$ a.e.

Let $\{z_\nu\}_{\nu=1}^\infty$ be a sequence of distinct points in Δ and let $\{a_\nu\}$ be a bounded sequence of complex numbers. Our remarks concern the interpolation problem

$$f(z_\nu) = a_\nu, \quad \nu = 1, 2, \cdots \tag{1.1}$$

for $f \in H^\infty$.

In §2 and §3 we give a new proof of Nevanlinna's Theorem [9] providing interpolating functions of constant modulus. At the same time we also prove a refinement of that theorem, due to Adamyan, Arov and Krein [1]. Our approach, using duality and extremal problems, yields a simpler proof than those found in [1] and [9], our method applies in the more general setting of logmodular algebras, and our Theorem 3 below seems new. However, we do not obtain the parameterizations of interpolating functions given in [1] or [9]. Thus §2 and §3 should be viewed as part of a program to derive known results using linear extremal problems. In §2 we simplify a lemma due to V. P. Havin [8], [10].

In §4 and §5 we give an extension of Carleson's theorem [3] describing those $\{z_\nu\}$ for which every interpolation (1.1) is possible. When $\{z_\nu\}$ is not an interpolating sequence, we give sharp conditions on $|a_\nu|$ which make (1.1) possible.

I thank Donald Marshall for helpful conversations.

2. Nevanlinna's theorem, which deserves to be better known, is this:

Theorem 1 [9]: Assume there are two distinct functions of norm less than or equal to one in H^∞ which satisfy the interpolation (1.1). Then there is an inner function in H^∞ which also satisfies (1.1).

The hypothesis that more than one H^∞ function fulfills (1.1) implies that $\Sigma(1 - |z_\nu|) < \infty$, so that the corresponding Blaschke product $B(z)$ exists. This means that if $f_0 \in H^\infty$ is one solution of (1.1), then every other solution has the form

$$f = f_0 + Bg, \quad g \in H^\infty.$$

Because of this, Theorem 1 is a corollary of the following theorem, due to Adamyan, Arov, and Krein [1].

*Partially supported by NSF Grant MPS-74-7035.

Theorem 2: Let $F_0 \in L^\infty$. Assume the coset $F_0 + H^\infty$ of L^∞/H^∞ contains two distinct functions of norm less than or equal to one. Then the coset contains a function F such that $|F| = 1$ a.e.

To pass from Theorem 2 to Theorem 1 simply let $F_0 = \bar{B}f_0$ and recall that $|B|=1$ a.e.

We divide the proof of Theorem 2 into two cases, although the first case can be subsumed under the harder second case. The norm of our coset is

$$\|F_0 + H^\infty\| = \inf\{\|F_0 + g\|:\ g \in H^\infty\}.$$

Case 1: $\|F_0 + H^\infty\| < 1$. Before starting we might note that in this case the hypothesis of the theorem is trivially satisfied. The argument we now give was independently found by Jacqueline Detraz.

For $p = 1, \infty$ let H_0^p be the set of H^p functions vanishing at $z = 0$. We have the duality relations

$$(H^1)^* = L^\infty/H_0^\infty;\ (H_0^1)^* = L^\infty/H^\infty.$$

Consider the extremal problem

$$a = \sup\{|\int F \frac{d\theta}{2\pi}|:\ F \in F_0 + H^\infty,\ \|F\| \le 1\}.$$

By the weak-star compactness of H^∞, there exists an extremal function $F \in F_0 + H^\infty$ such that $\|F\| \le 1$, $|\int F d\theta/2\pi| = a$. We claim $|F| = 1$ a.e. Since H^∞ contains constants and F is extremal we have

(2.1) $$\inf\{\|F + g\|:\ g \in H_0^\infty\} = 1.$$

On the other hand

(2.2) $$\inf\{\|F + g\|:\ g \in H^\infty\} = \|F_0 + H^\infty\| < 1.$$

By duality these give

(2.1)' $$\sup\{|\int Fh \frac{d\theta}{2\pi}|:\ h \in H^1,\ \|h\|_1 \le 1\} = 1$$

(2.2)' $$\sup\{|\int Fh \frac{d\theta}{2\pi}|:\ h \in H_0^1,\ \|h\|_1 \le 1\} < 1.$$

Choose $\{h_n\}$ in H^1 such that $\|h_n\|_1 \le 1$ and

(2.3) $$|\int F\, h_n \frac{d\theta}{2\pi}| \to 1.$$

Suppose there were a measurable set E of positive measure such that $|F| < \lambda < 1$ on E. Then by (2.3) $\int_E |h_n|\, d\theta/2\pi \to 0$. Because the logarithm is concave this means

$$\int_E \log|h_n| \frac{d\theta}{2\pi} \to -\infty.$$

Since $\|h_n\|_1 \le 1$, we then have

$$\log|h_n(0)| \le \int \log|h_n|\, d\theta/2\pi \to -\infty,$$

and $h_n(0) \to 0$. Consequently, setting $k_n = h_n - h_n(0)$, we have

$$\int F \frac{k_n}{\|k_n\|_1} \frac{d\theta}{2\pi} \to 1$$

which contradicts (2.2)'. That concludes the proof in Case 1.

Before turning to Case 2, we note that further information can be obtained in Case 1.

Theorem 3: If $\|F_0 + H^\infty\| < 1$, there is $F \in F_0 + H^\infty$ and $h \in H^1$, $h \neq 0$ so that

$$Fh = |h| \quad \text{a.e.} \tag{2.4}$$

A proof of Theorem 3 using the F. and M. Riesz theorem for logmodular algebras can be found in [2], but we would like to present a more elementary proof.

Proof of Theorem 3: Let F and $\{h_n\}$ be as in the above discussion. We claim the sequence $\{h_n\}$ has a subsequence weakly convergent in L^1. Letting H be the weak limit, we have $H \in H^1$, $H(0) \neq 0$ and (2.3) yields (2.4) with $h = \alpha H$, α constant, $|\alpha| = 1$.

If $\{h_n\}$ has no weakly convergent subsequence, then by p. 292 of [4], there are measurable sets $E_n \subset \partial\Delta$ such that

$$|E_n| \to 0,$$

where $|S|$ denotes the measure of S, and such that

$$\left|\int_{E_n} h_n \frac{d\theta}{2\pi}\right| > \beta > 0, \tag{2.5}$$

at least for a subsequence of $\{h_n\}$.

Lemma: If $\{E_n\}$ is a sequence of measurable subsets of $\partial\Delta$ such that $|E_n| \to 0$, then there is a sequence g_n of H^∞ functions such that

(i) $\sup_{E_n} |g_n| \to 0$

(ii) $g_n(0) \to 1$

(iii) $|g_n| + |1 - g_n| \leq 1 + \epsilon_n$

where $\epsilon_n \to 0$ $(n \to \infty)$.

Assuming the lemma for a moment, let

$$H_n = \frac{g_n h_n}{1 + \epsilon_n}, \quad K_n = \frac{(1 - g_n)h_n}{1 + \epsilon_n}.$$

Then H_n, $K_n \in H^1$, and $\|H_n\|_1 + \|K_n\|_1 \leq 1$, by (iii). Since $\epsilon_n \to 0$, (2.3) implies

$$\|H_n\|_1 \left|\int F \frac{H_n}{\|H_n\|_1} \frac{d\theta}{2\pi}\right| + \|K_n\|_1 \left|\int F \frac{K_n}{\|K_n\|_1} \frac{d\theta}{2\pi}\right| \to 1.$$

By (2.5) and (i), $\|K_n\|_1 > \beta > 0$, so that

$$\left|\int F \frac{K_n}{\|K_n\|_1} \frac{d\theta}{2\pi}\right| \to 1.$$

On the other hand, $K_n(0) \to 0$ by (ii), and we again contradict (2.2)'.

Proof of Lemma: Let $A_n \to \infty$ so that $A_n|E_n| \to 0$. Let F_n be the Poisson integral of

$$A_n \chi_{E_n} + i\, A_n^* \chi_{E_n}$$

where $^*u$ denotes the conjugate function of u. The analytic function F_n has values in the right half plane. Also $F_n(0) = A_n|E_n|$, and $\operatorname{Re} F_n = A_n$ on E_n. Let $G_n = (1 + F_n)^{-1}$. Then G_n has values in the disc

(2.6) $$|G_n - 1/2| \le 1/2,$$

and $G_n(0) \to 1$, $\sup_{E_n} |G_n| \to 0$. The disc (2.6) is collapsed inside the ellipse defined by (iii) under the mapping $z \to z^\delta$, if $\delta > 0$ is small. Letting $\delta_n \to 0$ slowly, we see that $g_n = G_n^{\delta_n}$ satisfies (i), (ii), and (iii).

It is easy to see that this lemma implies a similar lemma due to V. P. Havin [8], [10].

3. Case 2: $1 = \|F_0 + H^\infty\|$.

We have two distinct functions F_1, F_2 in $F_0 + H^\infty$ such that $\|F_1\| = \|F_2\| = 1$. Let P_z denote the Poisson kernel for $z \in \Delta$. Since $F_1 \not\equiv F_2$ there is $z \in \Delta$ such that

$$\int F_1 P_z \frac{d\theta}{2\pi} \ne \int F_2 P_z \frac{d\theta}{2\pi}.$$

Using a Möbius transformation, we can assume $z = 0$, and rotating F_1 and F_2 we can assume

(3.1) $$\operatorname{Re} \int F_1 \frac{d\theta}{2\pi} \ne \operatorname{Re} \int F_2 \frac{d\theta}{2\pi}.$$

Since L^∞/H^∞ is the dual space of H_0^1, the elements $F \in F_0 + H^\infty$ of norm one are the norm-preserving extensions to L^∞ of the linear functional

$$H_0^1 \ni h \to \int h F_0 \frac{d\theta}{2\pi}.$$

By (3.1) this functional has two extensions whose real parts disagree at the function $1 \in L^\infty H_0^\infty$. The proof of the Hahn-Banach theorem, therefore, gives

(3.2)
$$\sup_{h \in H_0^1} \{\|1 + h\| - \operatorname{Re} \int F_0 h \frac{d\theta}{2\pi}\} = m <$$
$$\inf_{h \in H_0^1} \{\|1 + h\| - \operatorname{Re} \int F_0 h \frac{d\theta}{2\pi}\} = M.$$

We may take $F \in F_0 + H^\infty$, $\|F\| = 1$ so that $\operatorname{Re} \int F\, d\theta/2\pi = M$. Then (3.2) can be rewritten as

(3.3) $$\inf_{h \in H_0^1} \{\|1 + h\| + \operatorname{Re} \int F(1 + h) \frac{d\theta}{2\pi}\} \ge M - m > 0,$$

(3.4) $$\inf_{h \in H_0^1} \{\|1 + h\| - \operatorname{Re} \int F(1 + h) \frac{d\theta}{2\pi}\} = 0.$$

By (3.4) there are $\{h_n\}$ in H_0^1 so that

$$\int (|1 + h_n| - \text{Re } F(1 + h_n))\frac{d\theta}{2\pi} \to 0. \tag{3.5}$$

Assume there is a measurable set $E \subset \partial\Delta$ of positive measure such that $|F| < \lambda < 1$ on E. Then

$$\int_E |1 + h_n| \frac{d\theta}{2\pi} \to 0 \tag{3.6}$$

by (3.5).

<u>Lemma</u>. If $E \subset \partial\Delta$ is a measurable set of positive measure, then there is $g \in H^\infty$ such that $g(0) = 1$ and such that g is real and negative on $(\partial\Delta)\setminus E$.

Assuming the lemma, we write $g \cdot (1 + h_n) = 1 + k_n$, $k_n \in H_0^1$. By (3.6)

$$\int_E (|1 + k_n| + \text{Re } F(1 + k_n)) \frac{d\theta}{2\pi} \to 0,$$

while

$$\int_{T\setminus E} (|1 + k_n| + \text{Re } F(1 + k_n)) \frac{d\theta}{2\pi} =$$

$$\int_{T\setminus E} |g|\{|1 + h_n| - \text{Re } F(1 + h_n)\} \frac{d\theta}{2\pi} \to 0$$

by (3.5). Hence

$$\int (|1 + k_n| + \text{Re } F(1 + k_n)) \frac{d\theta}{2\pi} \to 0,$$

contradicting (3.3), and $|F| = 1$ a.e.

<u>Proof of Lemma</u>. Using the characteristic function of E, we can construct an outer function G such that $1 \le |G| \le e$, $|G| = 1$ on $(\partial\Delta)\setminus E$, $|G| = e$ on E and $G(0) = R > 1$. The annulus $\{1 < |z| < e\}$ can be conformally mapped onto a domain bounded by the slit $[-4, 0]$ and an ellipse using the mapping

$$\varphi(z) = z + \frac{1}{z} - 2.$$

We then take

$$g = \frac{\varphi \circ G}{\varphi(R)}.$$

<u>Example</u>. When $\|F_0 + H^\infty\| = 1$ and the coset has a unique element of minimal norm, it may contain no unimodular function. Examples are easy to construct even in the setting of Nevanlinna's Theorem: Let the zeros z_ν of $B(z)$ accumulate at $z = 1$ and let $f_n \in (H^\infty)^{-1}$ satisfy $\|f_0\| \le 1$, $|f_0| \not\equiv 1$ but $|f_0| = 1$ on an arc I containing $z = 1$. We claim that if $g \in H^\infty$ and $\|f_0 - Bg\| \le 1$, then $g = 0$. Indeed,

$$|1 - Bgf_0^{-1}| \le 1,$$

on I, so that $f = Bgf_0^{-1}$ has positive real part on I. As is well known, this means the inner factor of f is analytic across I (set $v(\theta) = \chi_I(\theta) \arg f(e^{i\theta})$, so that $|v| \le \pi/2$. Then $f = e^{-i(v + i\tilde{v})}f$ is in $H^{1/2}$ and positive on I. Hence F and its inner factor reflect across I. See, for instance, [7].) Consequently $g = 0$.

4. Let $\{z_\nu\}$ be our sequence in Δ and write

$$\delta_\nu = \prod_{\mu,\mu\neq\nu} \left|\frac{z_\nu - z_\mu}{1 - \bar{z}_\mu z_\nu}\right|.$$

If $\inf_\nu \delta_\nu = \delta > 0$, then by Carleson's Theorem [3], every interpolation (1.1) has a solution $f \in H^\infty$ such that

(4.1) $$\|f\| \leq C \frac{1}{\delta} \log \frac{1}{\delta} \sup_\nu |a_\nu|,$$

where C is some constant. The dependence on δ in (4.1) is known to be sharp.

We want to consider (1.1) when $\inf_\nu \delta_\nu = 0$. We assume $\Sigma(1 - |z_\nu|) < \infty$, so that $\delta_\nu > 0$ for each ν. A. M. Gleason has observed (unpublished) that Earl's proof [6] of Carleson's theorem yields (1.1) whenever $|a_\nu| \leq \delta_\nu^2$. We give sharp conditions on the relation between $|a_\nu|$ and δ_ν which ensure the interpolation (1.1).

Let A(t) be some positive decreasing function on $[0, \infty)$.

<u>Theorem 4</u>: There is a constant C such that, if

$$\int_0^\infty A(t)dt < \infty$$

and if

(4.2) $$|a_\nu| \leq \delta_\nu A(1 + \log 1/\delta_\nu),$$

then the interpolation (1.1) has a solution $f \in H^\infty$ with

$$\|f\| \leq C \int_0^\infty A(t)\, dt.$$

On the other hand, if

$$\int_0^\infty A(t)\, dt = \infty$$

then there is a sequence $\{z_\nu\}$ and a sequence $\{a_\nu\}$ such that (4.2) holds but such that (1.1) has no solution in H^∞.

For example, interpolation is possible with a uniform bound on $\|f\|$ if we have

$$|a_\nu| \leq \delta_\nu(1 + \log 1/\delta_\nu)^{-2},$$

but interpolation is sometimes impossible if

$$|a_\nu| = \delta_\nu(1 + \log 1/\delta_\nu)^{-1}.$$

When $\inf_\nu \delta_\nu = \delta > 0$, Theorem 4 contains Carleson's theorem, including the bound (4.1). One sees this by letting A(t) be the characteristic function of $[0, 1 + \log 1/\delta]$. The relation between the two theorems is not surprising because the proofs are very similar.

The first part of the theorem can, of course, be phrased without reference to a function A(t), simply by taking the least decreasing function A(t) such that (4.2) holds. This gives the

<u>Corollary</u>: If

$$A_0 = \sum_{n=0}^{\infty} \sup_{\delta_\nu \leq e^{-n}} (|a_\nu|/\delta_\nu) < \infty,$$

then (1.1) has a solution $f \in H^\infty$ such that $\|f\| \leq C\, A_0$.

We give the proof in the upper half plane, where $z_\nu = x_\nu + y_\nu$, $y_\nu > 0$ and

$$\delta_\nu = \prod_{\mu,\mu\neq\nu} \left|\frac{z_\nu - z_\mu}{z_\nu - \bar{z}_\mu}\right|.$$

Assume

$$\int_0^\infty A(t)\, dt < \infty$$

and fix $\{a_\nu\}$ satisfying (4.2). For n fixed let

$$M_n = \inf\{\|f\|_\infty : \; f \in H^\infty,\ f(z_\nu) = a_\nu,\ 1 \leq \nu \leq n\}.$$

By normal families, the proof will be completed when we bound M_n by the right side of (4.3).

Let

$$B_n(z) = \prod_{\nu=1}^{n} \frac{z - z_\nu}{z - \bar{z}_\nu},$$

and let H^1 be the Hardy class in the upper half plane. By duality [3], we have

$$M_n = 2\pi \sup \{|\sum_{\nu=1}^{n} \frac{a_\nu h(z_\nu)}{B_n'(z_\nu)}| : \; h \in H^1,\ \|h\|_1 \leq 1\}$$

$$\leq 4\pi \sup \{ \sum_{\nu=1}^{n} y_\nu A(1 + \log 1/\delta_\nu)|h(z_\nu)| : \; h \in H^1,\ \|h\|_1 \leq 1\}$$

because $|B_n'(z_\nu)| \geq \delta_\nu/2y_\nu$.

A positive measure σ on the upper half plane is called a Carleson measure if $\int |h|\, d\sigma \leq C_1\|h\|_1$ for all $h \in H^1$ [5]. Carleson measures are characterized by the condition:

(4.4) $$\sigma(Q) \leq A_1 h$$

for all squares $Q = \{x_0 < x < x_0 + h,\ 0 < y < h\}$. Thus our job is to prove that

$$\sigma = \sum_{\nu=1}^{\infty} y_\nu A(1 + \log 1/\delta_\nu)\delta_{z_\nu}$$

satisfies (4.4) with

$$A_1 \leq \int_0^\infty A(t)\, dt,$$

where δ_z denotes the point mass at z.

Fix a square Q_0, with boundary on $y = 0$, of side h and let $T(Q_0)$ denote its top half. Partition $Q\backslash T(Q_0)$ into two squares Q_1 of side h/2 with disjoint interiors. Partition each $Q_1\backslash T(Q_1)$ into two squares Q_2 of side h/4 with disjoint interiors. Continuing, at stage m we have 2^m pairwise disjoint squares Q_m of side $2^{-m}h$. The

$T(Q_j)$, $j = 0, 1, \cdots$ form a pairwise disjoint covering of Q_0.

Considering only those $z_\nu \in Q_0$, write $z_\mu < z_\nu$ if there is a square Q_m in our decomposition of Q_0 such that

$$z_\nu \in T(Q_m), \quad z_\mu \in Q_m.$$

Roughly speaking this means z_μ is below z_ν. The case $z_\mu < z_\nu$ is included as a minor convenience.

Since $2 \log 1/t > 1 - t^2$, $t > 0$, we have

$$1 + \log 1/\delta_\nu \geq 1 + \frac{1}{2} \sum_{\mu, \mu \neq \nu} \{1 - \left|\frac{z_\nu - z_\mu}{z_\nu - \bar{z}_\mu}\right|^2\}$$

$$= 1 + \frac{1}{2} \sum_{\mu, \mu \neq \nu} \frac{4 y_\mu y_\nu}{|z_\nu - \bar{z}_\mu|^2} \geq \sum_{z_\mu < z_\nu} \frac{y_\mu}{y_\nu} = Y_\nu .$$

Since $A(t)$ decreases it suffices to prove

$$\sum_{z_\nu \in Q} y_\nu A(Y_\nu) \leq C_3 h \int_0^\infty A(t)\, dt. \tag{4.5}$$

For $m = 1, 2, \cdots$ let

$$E_m = \{z_\nu \in Q_0 : \ m \leq Y_\nu < m + 1\}.$$

We claim

$$\sum_{k=1}^{m} \sum_{E_k} y_\nu \leq 2(m + 1)h. \tag{4.6}$$

Let us assume (4.6) for a moment, finish the proof, and then come back to (4.6) itself. We have

$$\sum_{z_\nu \in Q} y_\nu A(Y_\nu) \leq \sum_{m=1}^{\infty} A(m) \sum_{E_m} y_\nu .$$

Summing the right side by parts twice and using (4.6), this yeidls

$$\sum_{z_\nu \in Q} y_\nu A(Y_\nu) \leq \sum_{m=1}^{\infty} (A(m) - A(m + 1))(m + 1)h$$

$$\leq C_3 h \int_0^\infty A(t)\, dt,$$

and (4.5) holds.

Proof of (4.6): Fix m and let $Q^1, Q^2, \cdots$ be those squares in our decomposition of Q_0 for which

$$E_k \cap T(Q_j) \neq \emptyset$$

for some $k \leq m$ and for which Q_j is maximal. The Q_j so chosen have disjoint interiors, and they cover $\bigcup_1^m E_n$. Fix Q_j and $z_\nu \in T(Q_j) \cap \bigcup_1^m E_k$. Then

$$\Sigma \{y_\mu : \ z_\mu \in Q_j \cap \bigcup_1^m E_k\} \leq y_\nu Y_\nu \leq 2(m + 1)\ell(Q_j)$$

where $\ell(Q_j)$ is the side length of Q_j. Summing over Q_j now gives (4.6).

5. We will only briefly show how to construct a counterexample when

$$\int_0^\infty A(t)\, dt = \infty.$$

Fix a square Q and fix an integer m. Place a point z_ν at the upper left corner of $T(Q_j)$ for each Q_j, $j = 0, 1, \cdots, m$. Then

$$1 + \log 1/\delta_\nu \leq c_1 Y_\nu,$$

and the inequalities opposite to (4.5) and (4.6) hold for this finite sequence of $N = 2^{m+1} - 1$ points. We can then choose $a_1, \cdots, a_N$ so that

$$|a_\nu| = \delta_\nu A(1 + \log 1/\delta_\nu)$$

and so that

$$M_N = 4\pi \sup \{\sum_1^N y_\nu A(1 + \log 1/\delta_\nu)|h(z_\nu)| : h \in H^1, \|h\|_1 \leq 1\}.$$

By the converse of the theorem on Carleson measures this means

$$M_N \geq c \int_1^m A(t)\, dt. \tag{5.1}$$

Therefore, Theorem 4 is sharp for finite sequences.

To obtain an infinite sequence for which (4.2) holds, but for which interpolation is impossible, choose a sequence $\{Q_m\}$ of squares of side one with lower left corner at say $x = 4^m$. In each Q_m do the above construction with m levels. Different squares Q_m do not interfere with each other, and (5.1) then holds for infinitely many m.

REFERENCES

1. V. M. Adamyan, D. Z. Arov, M. G. Krein, Infinite Hankel matrices and generalized Caratheodory-Fejer and I. Schur problems, Funct. Anal. Appl. 2 (1968), 269-281.
2. A. Bernard, J. B. Garnett, and D. E. Marshall, Algebras generated by inner functions, J. Func. Anal., (to appear).
3. L. Carleson, An interpolation problem for bounded analytic functions, Amer. Jour. Math. 80 (1958). 921-930.
4. N. Dunford, and J. T. Schwartz, Linear Operators, Part I. Interscience, New York, 1958.
5. P. L. Duren, Theory of H^p Spaces. Academic Press, New York, 1970.
6. J. P. Earl, On the interpolation of bounded sequences by bounded functions, Jour. London Math. Soc. (2) 2 (1970), 544-548.
7. T. W. Gamelin, J. B. Garnett, L. A. Rubel, and A. L. Shields, On badly approximable functions, J. Approx. Theory (to appear).
8. V. P. Havin, Weak completeness of the space L^1/H^1_0, Vest. Leningrad Univ., 13 (1973), 77-81 (Russian).
9. R. Nevanlinna, Über beschränkte analytische Funktionen, Ann. Acad. Sci. Fenn., Ser. A, 32 (1929), No. 7.
10. A. Pelczynski, Lectures, this conference.

NORM ATTAINING OPERATORS ON C(S) SPACES[1]

Jerry Johnson and John Wolfe
Oklahoma State University
Stillwater, OK 74074/USA

By "operator" we mean a bounded linear map between two Banach spaces. C(S) will denote the Banach space of continuous functions on the compact Hausdorff space S. An operator A is called norm attaining if $\|Ax\| = \|A\|$ for some x of norm one.

<u>Theorem 1</u>. If S and T are compact Hausdorff spaces, the norm attaining operators from C(S) to C(T) are norm dense among all operators from C(S) to C(T).

<u>Theorem 2</u>. If X and Y are Banach spaces, one of which is C(S), the finite rank norm attaining operators are norm dense among all compact operators from X to Y.

Theorem 2 answers a question in [1, page 6] while Theorem 1 was motivated by the question (posed in [2]) whether every C(S) space has what is called in [2] "property B".

<u>Proof of Theorem 1</u>. For the sake of brevity we prove the theorem for T totally disconnected. The heart of the more general proof is here without extra technical complications. Let A: $C(T) \to C(S)$ and $\varepsilon > 0$ be given. Define μ: $S \to C(T)^*$ by $\int f\, d\mu(s) = (Af)(s)$ for each $f \in C(T)$ and $s \in S$. Then μ is continuous for the weak*-topology on $C(T)^*$. Choose $s_0 \in S$ and $h \in C(T)$ such that $\int h\, d\mu(s_0) \geqq \|\mu\| - \varepsilon$, where $\|\mu\| = \sup\{\|\mu(s)\|: s \in S\}$. Since T is totally disconnected, h may be chosen with $|h(t)| = 1$ for each $t \in T$. Let $2/3 < r < 1$. We will construct a function μ': $S \to C(T)^*$ continuous for the weak*-topology on $C(T)^*$ such that (a) $\|\mu(s) - \mu'(s)\| \leqq r\varepsilon$ for each $s \in S$ and (b) $\sup_{s\in S} \|\mu'(s)\| \leqq r\varepsilon + \int h\, d\mu'(s_1)$ for some $s_1 \in S$.

Assuming this is done, we apply the construction inductively to obtain a sequence $\{\mu_n\}$ of weak* continuous mappings from S to $C(T)^*$ and points $\{s_n\}$ so that $\sup_{s\in S} \|\mu_n(s)\| \leqq r^n\varepsilon + \int h\, d\mu_n(s_n)$ and $\|\mu_n - \mu_{n+1}\| \leqq r^n\varepsilon$. Let A_n: $C(T) \to C(S)$ be defined by $A_n f(s) = \int f\, d\mu_n(s)$. Then $\|A_n\| \leqq r^n\varepsilon + A_n h(s_n)$ and $\|A_n - A_{n+1}\| \leqq r^n\varepsilon$. Together, these inequalities imply that $A' = \lim A_n$ exists, $\|A - A'\| \leqq \frac{r}{1-r}\varepsilon$, and $\|A'h\| = \|A'\|$, which is the desired conclusion.

To find s_1 and μ' satisfying (a) and (b), choose $\delta > 0$ with $2/3\, \varepsilon + \delta/2 < r\varepsilon$, and let $U = \{s \in S \mid \int h\, d\mu(s) > \|\mu\| - \varepsilon - \delta\}$. Put $M = \sup\{\|\mu(s)\|: s \in U\}$ and consider first the case $M \leqq \|\mu\| - \varepsilon/3$. Choose $\phi \in C(S)$ with $\phi(s_0) = h(s_0)$, $\phi = 0$ on $S\setminus U$ and $|\phi| \leqq 1$. Then $\mu'(s) = \mu(s) + \varepsilon/3\, \phi(s)\delta_{s_0}$ satisfies (a) and (b), where δ_{s_0} is the point mass at s_0 and $s_1 = s_0$. Now, suppose $M > \|\mu\| - \varepsilon/3$. Then there is a point $s_1 \in U$ such that $\|\mu(s_1)\| > \|\mu\| - \varepsilon/3$. Let $H^+ = \{t: h(t) = 1\}$ and $H^- = T\setminus H^+$, let (E^+, E^-) be a Hahn decomposition for $\mu(s_1)$, and let $A^+ = H^+ \cap E^+$ and $A^- = H^- \cap E^-$.

[1] We thank Joe Diestel, all the organizers and NSF for a truly great conference. An expanded version of this note will appear elsewhere.

Then $h + \chi_{E^+} - \chi_{E^-} = 2(\chi_{A^+} - \chi_{A^-})$. Hence, $\int [\chi_{A^+} - \chi_{A^-}]\, d\mu\,(s_1) = \frac{1}{2}\int h\, d\mu(s_1) + \frac{1}{2}\int [\chi_{E^+} - \chi_{E^-}]\, d\mu(s_1) \geq \frac{1}{2}(\|\mu\| - \epsilon - \delta) + \frac{1}{2}\|\mu(s_1)\| \geq \|\mu\| - (2/3\,\epsilon + \delta/2) > \|\mu\| - r\epsilon$. Hence, there are clopen sets $K^+ \subset H^+$ and $K^- \subset H^-$ such that $\int [\chi_{K^+} - \chi_{K^-}] d\mu(s_1) > \|\mu\| - r\epsilon$. Now $g = \chi_{K^+} - \chi_{K^-}$ belongs to $C(T)$ so there is a neighborhood W of s_1 such that $\int g\, d\mu(s) > \|\mu\| - r\epsilon$ for $s \in W$. Choose $\phi \in C(S)$ with $0 \leq \phi \leq 1$, $\phi(s_1) = 1$, and $\phi = 0$ on $S\backslash W$. Define $\mu'(s) = [1 - \phi(s)\, \chi_{S\backslash K}]\, \mu(s)$, where $K = K^+ \cup K^-$. First, $\|\mu'(s)\| \leq \|\mu(s)\|$ for every s since $0 \leq \phi(s)\, \chi_{S\backslash K} \leq 1$. Also, note that $\chi_{S\backslash K} \in C(T)$ so μ' is weak* continuous. Since $\mu'(s_1) = \chi_K \mu(s_1)$ and $h = 1$ on K^+ and $h = -1$ on K^-, we get $\int h\, d\mu'(s_1) = \int g\, d\mu(s_1) > \|\mu\| - r\epsilon$. Finally, $\mu'(s) = \mu(s)$ for $s \in S\backslash W$, and if $s \in W$, $\|\mu(s) - \mu'(s)\| \leq \|\chi_{S\backslash K}\, \mu(s)\| = \|\mu(s)\| - |\mu(s)|K \leq \|\mu\| - \int [\chi_{K^+} - \chi_{K^-}]\, d\mu(s) < r\epsilon$. This completes the proof of Theorem 1.

We will prove here only the half of Theorem 2 that answers the question posed in [1, page 6], namely, the case $X = C(S)$.

Given $A: \ C(S) \to Y$ of finite rank and $\epsilon > 0$, represent A by $Af = \sum_{j=1}^{n} \mu_j(f) y_j$, with $\sum_{j=1}^{n} \|y_j\| \leq 1$. Put $\mu = \sum_{j=1}^{n} |\mu_j|$ and $g_j - d\mu_j/d\mu$. There are simple functions h_j such that $\int |g_j - h_j| d\mu \leq \epsilon/2$ for each j. There is a collection of disjoint Borel sets $B_1, \cdots, B_m$ of positive μ-measure and there are numbers α_{ij}, $1 \leq i \leq n$, $1 \leq j \leq m$, so that $h_i = \sum_{j=1}^{m} \alpha_{ij}\, \chi_{B_j}$ for each i. Choose compact sets $K_j \subset B_j$ with $\mu(B_j \backslash K_j) < \epsilon/2m \max |\alpha_{ij}|$.

There exist disjointly supported continuous functions ϕ_j, with $\phi_j = 1$ on K_j and $0 \leq \phi_j \leq 1$. Define

$$Pf = \sum_{j=1}^{m} \left(\frac{1}{\mu K_j} \int_{K_j} f d\mu\right)\phi_j, \quad f \in C(S).$$

P is a projection with $\|P\| = 1$. The unit ball U of span $\{\phi_1, \cdots, \phi_m\}$ is compact so there is $u \in U$ with $\|APu\| = \sup\{\|APf\|: \ f \in U\}$. But this equals $\|AP\|$ since $U = \{Pf: \ f \in C(S), \|f\| \leq 1\}$. Thus, AP is norm attaining. Since $\int_{K_j} Pf d\mu = \int_{K_j} f d\mu$ for each j and f, it follows that

$$\int (Pf - f) \sum_j \alpha_{ij}\, \chi_{K_j}\, d\mu = 0 \text{ for each } f.$$

Now by construction $\int |\sum_j \alpha_{ij}\, \chi_{K_j} - h_i|\, d\mu < \epsilon/2$ and $\int |g_i - h| d\mu < \epsilon/2$ so $\epsilon > |\int (Pf - f)\, g_i\, d\mu| = |\int (Pf - f)\, d\mu_i|$. Hence,

$$\|APf - Af\| = \|\sum_{i=1}^{n} \mu_i(Pf)\, y_i - \mu_i(f)\, y_i\| \leq \sum_{i=1}^{n} |\int (Pf - f)\, d\mu_i|\, \|y_i\| \leq \epsilon$$

since $\sum_{i=1}^{n} \|y_i\| \leqq 1$. This completes the proof.

Many open questions about norm attaining operators have been posed in [1], [2] and [3]. We state here only two problems. (1) Take R^n with the usual inner product norm and let X be any Banach space. Are the norm attaining operators from X into R^n dense? In particular, how about $n = 2$?

(2) Are the norm attaining operators between the following pairs of spaces dense? (a) $C[0, 1]$ to L^1, (b) L^1 to $C[0, 1]$, (c) $C[0, 1]$ to ℓ^2, and (d) (asked by Uhl in [3]) L^1 to L^1.

REFERENCES

1. J. Diestel and J. J. Uhl, The Radon-Nikodym theorem for Banach space valued measures, Rocky Mountain Jour. Math., 6 (1976), 1-46.
2. J. Lindenstrauss, On operators which attain their norm, Israel Jour. Math., 1 (1963), 139-148.
3. J. J. Uhl, Norm attaining operators on $L^1[0, 1]$ and the Radon-Nikodym property, Pac. Jour. Math., 63 (1976), 293-300.

LOCAL UNCONDITIONAL STRUCTURE IN BANACH SPACES

H. Elton Lacey *
The University of Texas at Austin
Austin, TX 78712/USA

Introduction. The concept of local unconditional structure (or LUST) is an important tool in the study of the structure theory of Banach spaces. For example, in [14] Pelczyński uses it as a basis for the comparison of the classical spaces A and $C(\Omega)$, and in [17] Retherford discusses its relationship to ideals of operators. This paper provides an introduction to the general methods and techniques involved in LUST.

Throughout spaces can be either real or complex Banach spaces unless otherwise stated. The theory of Banach lattices will be assumed and standard Banach space terminology will be used without explanation.

Recall that if X is an n-dimensional Banach space, a basis $\{x_1, \cdots, x_n\}$ for X is said to be λ-unconditional, where $\lambda \geq 1$, if $\|\sum_{i=1}^{n} a_i x_i\| \leq \lambda \|\sum_{i=1}^{n} b_i x_i\|$ whenever $|a_i| \leq |b_i|$ for $i = 1, \cdots, n$. It is said to be unconditionally monotone if it is 1-unconditional. Moreover, it is easy to see that X admits an unconditionally monotone basis if and only if it admits a lattice structure under which it is a Banach lattice. The basic motivation for the notion of LUST is the observation that certain classical Banach spaces admit a family of finite dimensional subspaces which is "dense" in the family of all finite dimensional subspaces and such that each of the subspaces has a basis which is unconditional with respect to the same constant λ. Thus, for example, if $X = L_p(\mu)$, then such a family is the family of all subspaces spanned by finitely many disjoint integrable characteristic functions. Clearly these subspaces admit unconditionally monotone bases (and are, in fact, linearly isometric to $\ell_p(k)$ for some k) and their union is dense in X. On the other hand, if $X = C(\Omega)$ where Ω is a compact Hausdorff space, then the family of all subspaces spanned by a finite partition of unity (such a space is linearly isometric to $\ell_\infty(k)$ for some k) is dense in the sense that each finite dimensional subspace of C(K) can be, for each $\epsilon > 0$, embedded into one of these subspaces via an isomorphism T such that $\|T\|\|T^{-1}\| < 1 + \epsilon$. These ideas were exploited by Lindenstrauss and Pelczynski in [10] and Lindenstrauss and Rosenthal in [11].

Let X be a Banach space and consider the following conditions on the finite dimensional subspaces of X.

I. There are $\lambda > 1$ and a net (i.e., an upwards directed family under set inclusion) (E_γ) of finite dimensional subspaces of X such that $X = \overline{\bigcup_\gamma E_\gamma}$ and each E_γ admits a λ-unconditional basis.

II. There is a $\lambda > 1$ such that for every finite set $\{x_1, \cdots, x_n\}$ in X and every $\epsilon > 0$ there is a finite dimensional subspace F of X which has a λ-unconditional basis and $d(x_i, F) < \epsilon$ for $i = 1, \cdots, n$.

*This work was partially supported by NSF Grant #76-07634.

III. There is a $\lambda > 1$ such that for each finite dimensional subspace E of X there is a finite dimensional subspace $F \supset E$, a Banach lattice V and an isomorphism T of F onto V such that $\|T\|\|T^{-1}\| \leq \lambda$.

The corresponding isometric versions of I, II, and III are obtained by replacing the condition "there is a $\lambda > 1$" with the condition "for all $\lambda > 1$". Clearly conditions I and III are equivalent and I implies II. The fact that II implies I comes from the following lemma (see [9, 168]). A space satisfying one of these conditions is said to have <u>LUST</u>.

<u>Lemma 1</u>. Let X be a Banach space and $\mathfrak{F}$ a family of finite dimensional subspaces such that for each finite set $\{x_1, \cdots, x_n\}$ in X and each $\epsilon > 0$ there is an $F \in \mathfrak{F}$ such that $d(x_i, F) < 1 + \epsilon$ for $i = 1, \cdots, n$. Then for each finite dimensional subspace $G \supset E$ such that $d(G, F) < 1 + \epsilon$. If, in addition, there is a $\lambda \geq 1$ such that each $F \in \mathfrak{F}$ is the range of a projection from X of norm at most λ, then for each $\epsilon > 0$, G can be chosen so that it is the range of a projection from X of norm at most $\lambda + \epsilon$.

<u>Remark</u>. If the unconditional bases of these conditions are λ-equivalent to those of the class $\{\ell_p(n)\}$ for some fixed $1 \leq p \leq \infty$, then the space X is an $_{p,\lambda}$ space. As mentioned above, condition II shows that $L_p(\mu)$ and $C(\Omega)$ are $_{p,\lambda}$ and $_{\infty,\lambda}$ spaces respectively for all $\lambda > 1$.

Some additional concepts related to LUST play a role in the development of the theory.

IV. There is a $\lambda > 1$ such that for each finite dimensional subspace E of X there is a finite dimensional Banach lattice V and operators $T: E \to V$ and $S: V \to X$ such that $ST|_E = \mathrm{id}$ and $\|S\|\|T\| \leq \lambda$.

It is clear that LUST implies condition IV by condition III above. The converse question remains open. This condition has become known as the GL condition (for Gordon and Lewis who studied it in [5]) or <u>GL-LUST</u>. It was used in the $L_p(\mu)$ setting by Pietsch [16] to characterize complemented subspaces of $L_p(\mu)$, $1 < p < \infty$.

In the sequel it will be seen how spaces with LUST can be nicely embedded into Banach lattices and how they are related to the following concept.

V. X is isomorphic to a complemented subspace of a Banach lattice (for lack of a better term, X will be said to be a <u>CBL</u> when it has this property). To see how this is related to LUST, the following theorem is needed.

<u>Theorem 1</u>. Any Banach lattice has isometric LUST.

First note that the isometric version of condition II is true for Banach lattices with order continuous norm (see [9, p. 7]). The general case follows from the principle of local reflexivity and the fact that if X is a Banach lattice, then X^{**} is a Banach lattice with order continuous norm.

Remark. The principle of local reflexivity in fact shows that if X^{**} has (isometric) LUST or GL-LUST, then so does X. The converse question will be discussed later.

Thus Theorem 2 shows that if X is a CBL, then X has GL-LUST. Other relationships between LUST, GL-LUST, and CBL will be discussed in the sequel.

1. Embedding Techniques. In this section various ways of embedding spaces with LUST or G-LUST "nicely" into a Banach lattice are developed and discussed.

The technique used here will be that of ultraproducts as developed by Dacunha-Castelle, and Krivine in [2].

Let $(Y_a)_{a \in A}$ be a family of Banach lattices where A is a directed set and U a free ultrafilter on A which contains all the sets $\{a \in A: \ a \geq a_0\}$ for all $a_0 \in A$. Put $Y_0 = \{(x_a): \ \sup\|y_a\| < \infty\}$. Then under pointwise operations Y_0 is a lattice and the semi-norm $\rho(y_a) = \lim_U \|y_a\|$ preserves the lattice structure, i.e., $|x_a| \leq |y_a|$ for all a implies $\rho(x_a) \leq \rho(y_a)$. If $N = \{(y_a): \ \rho(y_a) = 0\}$, then for each (y_a), $\|(y_a) + N\|_\infty = \rho(y_a)$, that is, the norm obtained from ρ is the quotient norm in $Y = Y_0/N$ of the sup norm in Y_0. Thus Y is a Banach lattice.

If each Y_a is an $L_p(\mu_a)$ space, then Y is an $L_p(\nu)$ space for some ν since the norm on Y is p-additive (see [9, 135]).

If X has GL-LUST, then there is a natural embedding of X into an ultraproduct of Banach lattices as follows. Since X has GL-LUST there is a $\lambda > 1$ such that $X = \cup E_\gamma$ where (E_γ) is a net of finite dimensional subspaces and there are finite dimensional Banach lattices Y_γ and operators $T_\gamma: \ E_\gamma \to Y_\gamma$ and $S_\gamma: \ Y_\gamma \to X$ such that $S_\gamma T_\gamma | E_\gamma = \text{id}$ and $\|S_\gamma\| = 1$, $\|T_\gamma\| \leq \lambda$. Let Y_0, U and ρ be as above for the family (Y_γ) where the direction on the γ's is the natural set inclusion of the E_γ's. Define $J_0: \ X \to Y_0$ by $J_0 x = (y_\gamma)$ where $y_\gamma = T_\gamma(x)$ of $x \in E_\gamma$ and $y_\gamma = 0$ otherwise.

Thus for any $x \in E_\gamma$, $\|x\| = \|S_\gamma T_\gamma x\| \leq \|S_\gamma\| \ \|T_\gamma\| \ \|x\| \leq \lambda \ \|x\|$. Thus J_0 yields an isomorphism of X into $Y = Y_0/N$ such that $\|x\| \leq \|Jx\| \leq \lambda\|x\|$ for all $x \in X$.

The above embedding and notation will be preserved in the theorems below.

Theorem 2. Let X be a Banach space. Then X has GL-LUST if and only if X^{**} is a CBL.

Proof. If X^{**} is a CBL, then X^{**} has GL-LUST and by the principle of local reflexivity, X has GL-LUST.

Suppose X has GL-LUST. Then using the above notation let $y = (y_\gamma) \in Y_0$ and define $P_\gamma y = J(S_\gamma y_\gamma)$. Then for $y^* \in Y^*$, $|y^* \ (P_\gamma y)| = |y^*(J(S_\gamma y_\gamma))| \leq \|y^*|\lambda\|y_\gamma\|$. Thus $P_0(y) = w^* - \lim_U P_\gamma(y)$ exists and $\|P_0(y)\| \leq \lambda \ \rho(y)$ and it follows that P_0 has a lifting $P: \ Y \to X^{**}$ such that $\|P\| \leq \lambda$. Moreover, $PJx = x$ for all $x \in X$. Let Q be the natural embedding of X^* into X^{***}. Then Q^*P^{**} is a mapping from Y^{**} into X^{**} and $J^{**}Q^*P^{**}$ is the identity on X^{**}. Hence $J^{**}(X^{**})$ is complemented in Y^{**} with the projection being $Q^*P^{**}J^{**} = T$ and $\|T\| \leq \lambda^2$.

Remarks. Clearly if X is reflexive, then X has GL-LUST if and only if it is a CBL.

If $1 < p < \infty$ and X has GL-LUST with respect to the class $\{\ell_p(k)\}$, then X is isomorphic to a complemented subspace of an $L_p(\mu)$ space. This is a well known result due to Pietsch [16].

Corollary. Let X be a Banach space. Then X has GL-LUST if and only if X^* is a CBL. This follows immediately from the fact that X^* is complemented in X^{***} and the fact that the dual of a Banach lattice is a Banach lattice.

It should be remarked that since any Banach space X can be isometrically embedded into $C(\Omega)$ for some compact Hausdorff space Ω, the types of embeddings of spaces with GL-LUST or LUST must yield new information as in Theorem 1 to be of interest. Since any Banach lattice Y has LUST, if X is "nicely" embedded into Y, then X should have LUST. Such a concept of "nice" is now explored.

Definition 1. Let Y be a Banach lattice. A subspace Z of Y is said to be locally invariant if there is a $\lambda > 1$ such that for each finite dimensional subspace E of Y there is an isomorphism T of E into Z such that $\|T\|\,\|T^{-1}\| < \lambda$ and $T|E \cap Z = \mathrm{id}$.

Theorem 3. Let X be a Banach space. Then X has LUST if and only if it is isomorphic to locally invariant subspaces of a Banach lattice.

Proof. Since a Banach lattice has LUST, it follows immediately that any locally invariant subspace has LUST. To prove the converse we will use the notation of Theorem 1, putting $S_\gamma = T_\gamma^{-1}$. The isomorphism J constructed in Theorem 1 maps X into a locally invariant subspace of Y.

To see this let $Q_\gamma\colon Y_0 \to E_\gamma$ be the natural projection, i.e., $Q_\gamma(y_\delta) = y_\gamma$ for all $(y_\delta) \in Y_0$. Then if E is a finite dimensional subspace of Y and $z_i = (y_{\gamma,i})$ are such that $\{z_1 + N, \cdots, z_n + N\}$ forms a basis for E and $E' = \mathrm{span}\,\{z_i\}$, then a standard compactness argument shows that for each $\epsilon > 0$ there is a γ such that for $Q'_\gamma = Q_\gamma|E'$, $\|Q'_\gamma\|\,\|(Q'_\gamma)^{-1}\| < 1 + \epsilon$. Let $T_0 = J_0 T_\gamma^{-1} Q'_\gamma$. Then $\|T_0\|\,\|T_0^{-1}\| \leq (1 + \epsilon)\lambda^2$ and if $y \in E' \cap J_0(X)$ and $E_\gamma \supseteq J_0^{-1}(E' \cap J_0(X))$, then $T_0 y = y$. The operator $T\colon E \to Y$ defined by $T(y + N) = T_0(z) + N$ where $z \in E'$ and $y + N = z + N$ is well defined and an easy calculation shows that $\|T\|\,\|T^{-1}\| \leq (1 + \epsilon)\lambda^2$ and $T|E \cap J(X) = \mathrm{id}$.

2. Complemented Subspaces of Banach Lattices. One of the most interesting problems concerning LUST is whether or not every CBL has LUST. The answers to this and the question of whether or not both X and X^* have LUST whenever one of them does are both unknown. By results in section one it is clear that if every CBL has LUST, then both X and X^* have LUST whenever one of them does.

In this section a technique for proving that certain spaces which are CBL's have LUST is developed. This technique envelopes all of the known results of this type. It is based upon a careful analysis of the proof by Lindenstrauss and Rosenthal [10] that a complemented subspace of an $L_p(\mu)$ space whish is not isomorphic to a Hilbert space is an $\mathcal{L}_p$ space.

To introduce the idea the following definition is made. The reason for it will become apparent in the following theorems.

Definition 2. Let $X = Y + Z$. Then Z is said to be locally disjoint from Y if there are constants $K > 0$, $\eta > 0$ such that for each pair of finite dimensional subspaces E and F of Y and Z respectively, there is an isomorphism τ of F into Y such that $\|\tau\| \|\tau^{-1}\| \leq K$ and $\|y + \tau z\| \geq \eta\|y\|$ for all $y \in E$ and $z \in F$.

Remarks. 1. The proof used by Lindenstrauss and Rosenthal [10] shows that if $L_p(\mu) = Y + Z$ and Y is not a Hilbert space, then Z is locally disjoint in Y. The point is that Y, itself, contains a subspace Y_0 which is isomorphic to ℓ_p and complemented in $L_p(\mu)$. Hence it follows that given finite dimensional subspaces F of Z and E of Y, F can be placed inside Y_0 in such a way as to be "far away" from E in the above sense.

2. Another case was noticed by Figiel,Johnson, and Tzafriri [4]. Suppose Y contains $\ell_\infty(n)$'s uniformly. Then, in fact, for each $\varepsilon > 0$ and each n there is an isomorphism T of $\ell_\infty(n)$ into Y such that $\|T\| \|T^{-1}\| < 1 + \varepsilon$ [12]. Since every finite dimensional subspace F can be almost isometrically embedded into some $\ell_\infty(n)$, one can again obtain that Z is locally disjoint in Y. (Using only the uniform embedding instead of $1 + \varepsilon$, one can also show that Z is locally disjoint in Y by a simple calculation.)

Theorem 4. Let X be a Banach space with LUST. If $X = Y \oplus Z$ and Z is locally disjoint in Y, then Y has LUST.

Proof. Let P be the projection of X onto Y with kernel Z, $L = \|P\|$, K and of the constants of local disjointness and λ the constant of LUST. If E is a finite dimentional subspace of Y, then there is a finite dimensional subspace F of X such that $F \supset E$ and F admits a λ-unconditional basis. Now, $F \subset F_1 \oplus F_2$ where $F_1 \subset Y$ and $F_2 \subset Z$. Hence it suffices to find an isomorphism T of $F_1 \oplus F_2$ so that $\|T\|\|T^{-1}\|$ is bounded by a universal constant not depending on E, F, F_1 or F_2. By the local disjointness there is an isomorphism τ of F_2 into Y such that $\|\tau\| = 1$ and $\|\tau^{-1}\| \leq K$ and $\|y + \tau z\| \geq \eta\|y\|$ for all $y \in E$ and $z \in F_2$. Let $T(y + z) = y + \tau z$ for all $y \in F_1$ and $z \in F_2$. Then $\|T(y + z)\| \leq \|y\| + \|\tau\| \|z\| \leq \|y\| + \|z\| \leq 2L\|x + y\|$. On the other hand, $\|T(y + z)\| = \|y + \tau z\| \geq \eta\|y\|$ and $\|y + \tau z\| \geq \|\tau z\| - \|y\| \geq (1/\|\tau^{-1}\|)\|z\| - (1/\eta)$ $\|y + \tau z\|$. Hence, $\|y + \tau z\| \geq (\eta + (1/K)(\eta/1 + \eta)(\|y\| + \|z\|)$ and it follows that $\|T\| \|T^{-1}\|$ is bounded by a constant depending only on η, K and L.

Corollary. If X is a Banach lattice and $X = Y \oplus Z$ where Z is locally disjoint in Y, then Y has LUST.

The following theorem (proved by Stern in [19]) actually incorporates both the results in the above remarks.

Theorem 5. Let X be a Banach space and suppose $X = Y \oplus Z$. If X is finitely representable in Y, then Z is locally disjoint in Y.

Proof. The idea of the proof is as follows. If λ is the norm of projection of X onto Y with kernel Z, then for each $0 < \epsilon < 1$, each integer k, and each finite dimensional subspace F of Z there are k isomorphic copies F_i of F in Y such that $\|y + z\| \geq \lambda^{-1}(1 - \epsilon)\|y\|$ for all $y \in F_i$ and $z \in F_j$ with $i \neq j$. Suppose $k = 2$ and $\eta > 0$. There is an isomorphism τ_1 of F into Y such that $\|\tau_1\| = 1$ and $\|\tau_1^{-1}\| < 1 + \eta$. There is an isomorphism τ_2 of $F \oplus \tau_1(F)$ into Y such that $\|\tau_2\| = 1$ and $\|\tau_2^{-1}\| < 1 + \eta$. There is an isomorphism τ_2 of $F \oplus \tau_1(F)$ into Y such that $\|\tau_2\| = 1$ and $\|\tau_2^{-1}\| < 1 + \eta$. Now, $\|\tau_2 z + \tau_2\tau_1 w\| \geq (1/\|\tau_2^{-1}\|)\|z + \tau_1 w\| \geq (1/1 + \eta)\lambda^{-1}\|z\| \geq \lambda^{-1}(1/1 + \eta)\|\tau_2 z\|$. The result follows by choosing the appropriate $\eta > 0$. The extension to k follows by induction. Now put $\alpha = \lambda^{-1}(1 - \epsilon)$ where $0 < \epsilon < 1$ is fixed.

Let $\eta > 0$ (to be chosen more precisely later). If E is a finite dimensional subspace of Y, let k be an integer such that for any $y_1, \cdots, y_{k+1}$ in E of norm one, $\|y_i - y_j\| < \eta$ for some $i \neq j$. Let $F_1, \cdots, F_{k+1}$ be chosen as above, i.e., $\|z_i - z_j\| \geq \alpha\|z_i\|$ for all $z_i \in F_i$, $z_j \in F_j$, $i \neq j$. Then for some $1 \leq i \leq k + 1$, $\|y + z\| \geq \eta\alpha\|y\|$ for all $y \in E$, $z \in F_i$. For, if not, then there are $y_i \in E$ of norm one and $z_i \in F_i$ such that $\|y_i + z_i\| < \eta\alpha$ for $i = 1, \cdots, n$. But for some $i \neq j$, $\|y_i - y_j\| < \eta$. Hence $1 - \eta\alpha \leq \|z_i\| \leq \|z_i - z_j\| \leq \|z_i + y_i\| + \|-y_i + y_j\| + \|-y_i - z_j\| < \eta + 2\eta\alpha$.

Thus $1 < \eta + 3\eta\alpha$ which is not true for small enough η. Hence, it follows that Z is locally disjoint in Y.

Corollary. If X has LUST and $X = Y \oplus Z$ where X is finitely representable in Y, then Y has LUST. In particular, it is true if X is a Banach lattice.

Remarks. As remarked above, this theorem envelops all of the known results concerning when a CBL has LUST. To be more precise, if Y is a complemented subspace of $L_p(\mu)$ and Y is not isomorphic to a Hilbert space, then Y contains a subspace Y_0 isomorphic to ℓ_p. But Krivine [8] has shown that this implies that ℓ_p, and hence $L_p(\mu)$, is finitely represented in Y_0.

Also, as mentioned above, if Y is any Banach space which contains $\ell_\infty(n)$'s uniformly, then, in fact, every Banach space is finitely represented in Y. Thus, the following corollary is immediate.

Corollary. If X has GL-LUST, then $X \oplus c_0$ has LUST. In particular, X has GL-LUST if and only if it is isomorphic to a complemented subspace of a space with LUST.

It has already been observed that Hilbert space plays a role in LUST in L_p spaces, namely, the elegant result of Lindenstrauss and Pelczynski that a complemented subspace of $L_p(\mu)$ is either a Hilbert space of an $\mathcal{L}_p$ space.

Due to the famous result of Dvoretsky that ℓ_2 is finitely represented in any infinite dimensional Banach space, ℓ_2 also plays a role in LUST in general spaces. The following lemma is well known.

Lemma 2. Let X be a Banach space and E a finite dimensional subspace of X. Then for each $0 < \epsilon < 1$ there is a subspace $Y \supset E$, Y has finite codimension in X, and there is a projection of Y onto E of norm less than $1 + \epsilon$.

Proof. Let $\eta > 0$ and $x_1, \cdots, x_n$ be an η-net for the unit sphere of E. Choose $x_i^* \in X^*$ with $1 = \|x_i^*\| = x_i^*(x_i)$ and let $Z = \{x: x_i^*(x) = 0, i = 1, \cdots, n\}$. Then $Z \cap E = \{0\}$ and $Y = E + Z$ has finite codimension in X. If $x \in E$ is of norm one, then for any $z \in Z$, $\|x + z\| \geq |x_i^*(x)| \geq 1 - \eta$. Thus, for any $x \in E$, $z \in Z$, $\|x\| \leq (1/1 - \eta)\|x + z\|$ and by appropriate choice of η, the natural projection of Y onto E has norm less than $1 + \epsilon$.

Theorem 6. If $X \oplus \ell_2$ has LUST, then so does X.

Proof. Suppose X is infinite dimensional. If E is a finite dimensional subspace of X and $\epsilon > 0$, choose a subspace N as above such that the natural projection of $N \oplus E$ onto E has norm less than $1 + \epsilon$. Since ℓ_2 is finitely represented in N, it follows that ℓ_2 is locally disjoint in X.

An example of a space which does not have GL-LUST is now given. The technique is to show that for any quasireflexive non-reflexive space X (i.e., $1 \leq \dim X^{**}/X < \infty$), X^{**} is not a CBL. The ideas involved are from [4] and [7].

Definition 3. Let X be a Banach lattice and suppose P is a projection on X. Then X is said to be P-accessible if for each $x \in X$ and $\epsilon > 0$ there is a $y \in X$ such that $|y| \leq |x|$ and $\|Py\| > \|x\| - \epsilon$.
The reason for this definition will become apparent below.

Theorem 7. Let X be a Banach lattice and P a projection of X onto Y. If Y does not contain a subspace isomorphic to c_0 and X is P-accessible, then X does not contain a subspace isomorphic to c_0.

Proof. Suppose X contains a subspace isomorphic to c_0. Then it contains a sublattice isomorphic to c_0 (see [13]). Thus there is a disjoint positive sequence (x_n) in X K-equivalent to the unit vector basis of c_0. Since X is P-accessible there are $y_n \in X$ such that $|y_n| \leq x_n$ and $\|y_n\| > \frac{1}{2}\|x_n\|$ for all n. Hence, for any $a_1, \cdots, a_n$ it follows that $\|\sum_{i=1}^{n} a_i P(y_i)\| \leq \|P\|\|\sum_{i=1}^{n} a_i y_i\| \leq \|P\|\|\sum_{i=1}^{n} a_i x_i\| \leq \|P\|K \sup|a_i|$. This means that Y contains a subspace isomorphic to c_0 (see [15]).

To see how this applies a re-norming technique from [4] is needed. The idea is the following. Suppose X is a Banach lattice and P is a projection of X onto a subspace Y. Then there is a semi-norm ρ on X such that ρ is equivalent to the original norm on Y and the projection P is accessible with respect to the norm given by ρ. Let $\rho(x) = \sup\{\|Py\|: |y| \leq |x|\}$. Then clearly ρ is a semi-norm and if $|x_1| \leq |x_2|$, then $\rho(x_1) \leq \rho(x_2)$. Moreover, for any x, $\rho(x) \leq \|P\|\|x\|$ and if $y \in Y$, $\|y\| \leq \rho(y) \leq \|P\|\|y\|$. If $\rho(x) = 0$, then $Px = 0$ so that P lifts to a projection of

X/N onto Y, where $N = \{x: \rho(x) = 0\}$. Moreover, $\|P(x + N)\| = \rho(Px) \leq \|P\|\|Px\| \leq \|P\|\rho(x) = \|P\|\|x + N\|$ and the projection is bounded. Let $\epsilon > 0$ be given. Then for $x \in X$, there is a $y \in X$ such that $|y| \leq |x|$ and $\rho(y) \geq \|Py\| > \rho(x) - \epsilon$ and it follows that X is P-accessible.

Theorem 8. If X is quasi-reflexive and non-reflexive, then X does not have GL-LUST.

Proof. If X has GL-LUST, then X^{**} is a CBL and, hence, X is a CBL. In particular, by the above X is isomorphic to a subspace X_0 of a Banach lattice Y and there is a projection P of Y onto X_0 so that X is P-accessible. But, X_0 does not contain a subspace isomorphic to c_0 and, hence, the norm on Y is order continuous (see [13]). By a theorem of Trzafriri [20], it follows that X must contain a subspace isomorphic to ℓ_1, which is also impossible.

Remark. This is one of the simpler examples of spaces without GL-LUST. In [5], Lewis and Gordon prove that the space of compact linear operators from ℓ_2 to ℓ_2 does not have GL-LUST and in [14] Pelczynski observes that using results in [5] it follows that the space A of analytic functions on the unit disk does not have GL-LUST.

3. Duality and LUST. As mentioned above, the main problem of duality and LUST, namely, does both X and X^* have LUST whenever one of them does is still open. Of course, GL-LUST preserves duality as was proved in section one.

The following theorem was proved in [4].

Theorem 9. Let X be a Banach space. Then X has LUST if and only if X^{**} does. Another result concerning duality and LUST was recently noticed by J. Lindsay. The following theorem of Johnson [6] will be needed.

Theorem 10. Let X be a Banach space with LUST. Then every subspace of X is either superreflexive or contains $\ell_\infty(n)$'s uniformly or contains $\ell_1(n)$'s uniformly and uniformly complemented.

Theorem 11. Let X be a Banach space which is not superreflexive and has GL-LUST. Then either X or X^* has LUST.

Proof. If X contains $\ell_\infty(n)$'s uniformly, then since X^{**} is a CBL, X^{**} has LUST by Theorem 4 of section 2 and by the principle of local reflexivity, X has LUST. On the other hand, if X does not contain $\ell_\infty(n)$'s uniformly, then neither does X^{**} and by Theorem 7 X^{**} is embeddable in a Banach lattice Y which does not contain $\ell_\infty(n)$'s uniformly. Since Y has LUST, it follows from Johnson's theorem above that X contains $\ell_1(n)$'s uniformly and uniformly complemented so that X^* contains $\ell_\infty(n)$'s uniformly. Since X^* is a CBL, it has LUST by Theorem 5 of section 2.

4. A Local Characterization of Banach Lattices. Since Banach lattices play a large role in LUST, it seems reasonable that they can be characterized by their local un-

conditional structure. The first result along these lines was due to Zippin [22] where he showed that if $X = \bigcup_{\gamma} E_{\gamma}$ where (E_{γ}) is a net of finite dimensional subspaces each linearly isometric to some $\ell_p(k)$, then $X = L_p(\mu)$ for some measure μ, $1 \leq p < \infty$. This was extended to separable $\mathcal{L}_{p,\lambda}$ spaces for all $\lambda > 1$ by Lindenstrauss and Pelczynski in [10] and to the general case by Tzafriri in [21]. The proof involves the theory of contractive projections on an $L_p(\mu)$ space (see [9] for a systematic development of this).

Recently Bernau and Lacey [1] gave a characterization of Banach lattices with order continuous norm in terms of LUST. The proof uses the theory of LUST in a direct manner. It is illustrated for the case of $\mathcal{L}_p$ spaces below. (I am grateful to S. J. Bernau for letting me adapt this development from his seminar notes.)

<u>Theorem 12</u>. If X is an $\mathcal{L}_{p,\lambda}$ space for all $\lambda > 1$, then X is linearly isometric to $L_p(\mu)$ for some measure μ.

The proof will be a consequence of a sequence of lemmas. The first one is a form of a Clarkson inequality.

<u>Lemma 3</u>. Let $1 \leq p < \infty$ x, $y \in L_p(\mu)$. If $f(x, y) = 2\|x\|^p + 2\|y\|^p - \|x + y\|^p - \|x - y\|^p$, then $f(x, y) \geq 0$ if $p \leq 2$ and $f(x, y) \leq 0$ if $p \geq 2$. In addition, $2|2^{p/2} - 2|\||x| \wedge |y|\|^p \leq |f(x, y)|$.

<u>Proof</u>. Suppose z, w are complex numbers and $h(z, w) = 2|z|^p + 2|w|^p - |z + w|^p - |z - w|^p$. If $zw = 0$, then $h(z, w) = 0$. Assume $|z| = 1$ and $|w| = r \geq 1$. If $t = \arg z\bar{w}$, then $|z + w|^p + |z - w|^p = (1 + 2rt + r^2)^{p/2} + (1 - 2rt + r^2)^{p/2}$. For a fixed r this has the minimum (maximum) value $2(1 + r^2)^{p/2}$ if $p \geq 2$ $(p \leq 2)$. Now $2 + 2r^p - 2(1 + r^2)^{p/2}$ is a decreasing (increasing) function of r on $[1, \infty)$ if $p \geq 2$ $(p \leq 2)$ and has value $2(2 - 2^{p/2})$ at $r = 1$. The lemma follows by integration.

<u>Corollary</u>. $f(x, y) = 0$ if and only if $|x| \wedge |y| = 0$.

The following notation will be used throughout for finite dimensional Banach lattices of $L_p(\mu)$. For x, $y \in X$, $a(x, y) = ((\mathrm{Re}(x\,\overline{\mathrm{sgn}\,y})^+ \wedge |y|)\mathrm{sgn}\,y$. Clearly, the only thing to define is sgn y. If $X = L_p(\mu)$, then sgn y is the ordinary signature function of y. If X is a finite dimensional Banach lattice, then it has a unique positive normalized basis $x_1, \cdots, x_n$. If $x = \sum_{i=1}^{n} a_i x_i$, then $\bar{x} = \sum_{i=1}^{n} \bar{a}_i x_i$ and $\mathrm{sgn}\,x = \sum_{i=1}^{n} (\mathrm{sgn}\,a_i)x_i$.

<u>Lemma 4</u>. Let X be as above. Then for u, v, u', v' in X and $0 < \theta \leq 1$, then $\|a(u, v) - a(u', v')\| \leq 4/\theta(\|u - u'\| + \|v - v'\|) + 2\theta\|u\|$.

<u>Proof</u>. It is clearly sufficient to prove the inequality for complex numbers. If $|\mathrm{sgn}\,v - \mathrm{sgn}\,v'| > \theta$, then $|v - v'| = ||v|(\mathrm{sgn}\,v - \mathrm{sgn}\,v') + (|v| - |v'|)\mathrm{sgn}\,v'| > \theta|v| - |v - v'|$. Hence $|v| < 2|u - v'|\theta^{-1}$ and similarly for v'. Since $|a(u, v)| \leq |v|$, the result follows.

Now suppose that $|\operatorname{sgn} v - \operatorname{sgn} v'| \leq \theta$. Then $|a(u, v) - a(u', v')| \leq (\operatorname{Re} u \overline{\operatorname{sgn} v})^+ \wedge |v| \, |\operatorname{sgn} v - \operatorname{sgn} v'| + |(\operatorname{Re} u \overline{\operatorname{sgn} v})^+ \wedge |v| - (\operatorname{Re} u' \overline{\operatorname{sgn} v'})^+ \wedge |v'| \leq \theta|u| + |u \overline{\operatorname{sgn} v} - u' \overline{\operatorname{sgn} v'}| + |v - v'| \leq \theta|u| + \theta|u| + |u - u'| + |v - v'|$. Since $\theta < 1$, $4/\theta > 1$ and the result follows.

The next lemma is a sort of universal approximation lemma for finite dimensional Banach lattices.

Lemma 5. Let $0 < \eta < 1$. If E is a finite dimensional Banach lattice, then for any $x, y \in E$ there are an $N < 20\eta^{-4} - 1$, $\lambda_1, \cdots, \lambda_N$ with $|\lambda_i| \leq 1/\eta$ for $i = 1, \cdots, N$, and mutually disjoint elements $f_0, \cdots, f_N$ in E such that

$$\|x - f_0 - \sum_{i=1}^{N} \lambda_i f_i\| + \|y - \sum_{i=1}^{N} f_i\| \leq \eta(\|x\| + \|y\|).$$

Proof. The main observation is that a square in the complex plane with sides of length $2/\eta$ can be covered with not more than $[(2/\eta)/(\eta/\sqrt{2}) + 1]^2 = (1 + 2\sqrt{2}\,\eta^{-2})^2$ squares with sides of lengths at most $\eta/\sqrt{2}$. Thus the closed disc centered at the origin and of radius $1/\eta$ may be covered with N disjoint sets $H_1, \cdots, H_N$ of diameter at most η where $N + 1 < (1 + 2\sqrt{2}\,\eta^{-2})^{-} < 20\eta^{-4}$.

Now choose $\lambda_i \in H_i$ and suppose $e_1, \cdots, e_n$ is the positive normalized disjoint basis for E. Then $x = \sum_{i=1}^{n} \alpha_i e_i$ and $y = \sum_{i=1}^{n} \beta_i e_i$. Define

$$K_0 = \{i: |\beta_i| < |\alpha_i|\eta\}, \quad f_0 = \sum_{i \in K_0} a_i e_i, \quad K_j = \{i: \alpha_i / \beta_i \in H_j\}$$

and $f_i = \sum_{i \in K_j} \beta_i e_i$ for $j = 1, \cdots, N$. Then

$$\|x - f_0 - \sum_{i=1}^{N} \lambda_i f_i\| + \|y - \sum_{i=1}^{n} f_i\| = \|\sum_{i=1}^{N} \sum_{j \in K_i} (\alpha_j - \lambda_i \beta_j) e_j\| + \|\sum_{j \in K_0} \beta_j e_j\| \leq \eta(\|y\| + \|x\|).$$

The next lemma gives an estimate of how well isomorphisms in $L_p(\mu)$ spaces preserve disjointness. It follows immediately from Lemma 2.

Lemma 6. Let $1 \leq p < \infty$, $p \neq 2$. If T is an isomorphism of $L_p(\mu)$ into $L_p(\nu)$ with $\|T\| \leq \lambda$ and $\|T^{-1}\| = 1$, then for disjoint x, y in $L_p(\mu)$, $\||Tx| \wedge |Ty|\| \leq |2^{p/2} - 2|^{-1/p} (\lambda^p - 1)^{1/p} (\|x\| + \|y\|)$.

The next lemma is the main key to the theorem. It is an estimate of how well isomorphisms of finite dimensional L_p spaces preserve the operation $a(x, y)$.

Lemma 7. Let $1 \leq p < \infty$, $p \neq 2$, $E = \ell_p(k)$ for some integer k, T an isomorphism of E into $L_p(\nu)$, $\|T\| \leq \lambda$ and $\|T^{-1}\| = 1$. If $\lambda^p < 2$ and $\epsilon = (\lambda^p - 1)^{1/p}$, then there is a constant A independent of E, $L_p(\nu)$ and T such that $\|a(Tx, Ty) - T(a(x, y))\| \leq A\epsilon^{-19} (\|x\| + \|y\|)$ for all x and y in E.

Proof. Let θ, η be positive constants with $\eta < 1$ and $x, y \in E$. By lemma 5 choose $f_0, \cdots, f_N$ disjoint in E and scalars $\lambda_1, \cdots, \lambda_N$ such that $|\lambda_i| \leq 1/\eta$, $N + 1 < 20\eta^{-4}$, and $\|x - f_0 - \sum_{i=1}^{N} \lambda_i f_i\| + \|y - \sum_{i=1}^{N} f_i\| \le \eta(\|x\| + \|y\|)$. Now, put

$$x^* = f_0 - \sum_{i=1}^{N} \lambda_i f_i \text{ and } y^* = \sum_{i=1}^{N} f_i \text{ and } f_i^* = Tf_i - \sum_{j\neq i} (|Tf_i| \wedge |Tf_j|)\operatorname{sgn} Tf_i$$

and observe that $|f_i^*| \wedge |f_j^*| = 0$ if $i \neq j$. If $K = |2^{p/2} - 2|^{-1/p}$, then by Lemma 6, $\|Tf_i - f_i^*\| \le \sum_{j\neq i} \||Tf_i| \wedge |Tf_j|\| \le K\, N\, \epsilon(\|Tf_i\| + \|Tf_j\|) \le K\, N\lambda\, \epsilon\, (\|x\| + \|y\|)$. Moreover for $u^* = f_0^* + \sum_{i=1}^{N} \lambda_i f_i$ and $v^* = \sum_{i=1}^{N} f_i^*$, it follows that $\|Tx^* - u^*\| + \|Ty^* - v^*\| = \|Tf_0 - f_0^* + \sum_{i=1}^{N} \lambda_i\, (Tf_i - f_i^*)\| + \|\sum_{i=1}^{N} (Tf_i - f_i^*)\| \le (N + 1)K\, N\lambda\, \epsilon\, (\|x\| + \|y\|) + \eta^{-1} N\, K\, N\lambda\, \epsilon(\|x\| + \|y\|) \le \eta^{-1} 2K\, \lambda(N + 1)^2\, \epsilon(\|x\| + \|y\|)$. Also, $\|T(a(x^*, y^*)) - a(u^*, v^*)\| = \|\sum_{i=1}^{N} a(\lambda_i, 1)(Tf_i - f_i^*)\| \le N\, K\, N\lambda\, \epsilon(\|x\| + \|y\|)$. Hence by Lemma 4, $\|T(a(x, y)) - a(x^*, y^*)\| \le \|T\|\|a(x, y) - a(u^*, v^*)\| + KN^2\lambda\, \epsilon(\|x\| + \|y\|) \le \lambda[4\theta^{-1}\eta(\|x\| + \|y\|) + 2\theta(\|x\| + \|y\|) + KN^2\lambda\, \epsilon(\|x\| + \|y\|)] = (\lambda\, 4\theta^{-1}\eta + 2\theta + K\, N^2\epsilon)(\|x\| + \|y\|)$.

On the other hand, by Lemma 4 again, $\|a(Tx, Ty) - a(u^*, v^*)\| \le 4\theta^{-1}(\|Tx - u^*\| + \|Ty - v^*\|) + 2\theta\|Tx\| \le 4\theta^{-1}[\|T\|(\|x - x^*\| + \|y - y^*\|) + \|Tx^* - u^*\| + \|Ty^* - v^*\|] + 2\theta\|T\|(\|x\| + \|y\|) \le |4\theta^{-1}(\lambda\eta + \eta^{-1}\, 2K\, \lambda(N + 1)^2\epsilon) + 2\lambda\theta] \cdot (\|x\| + \|y\|)$.

Thus for $A = A(\lambda, \theta, \eta, \epsilon) = \lambda[8\eta\theta^{-1} + 4\theta + K\, N^2\epsilon + \eta^{-1}2K(N + 1)^2\epsilon] \le 2[8\eta\theta^{-1} + 4\theta + 4K(20/\eta^{-1})^2\eta^{-1}\epsilon] = B$ it follows that $\|a(Tx, Ty) - T(a(x, y))\| \le A(\|x\| + \|y\|)$. Moreover, putting $\eta = \theta^2$ it follows that $B = 2(12\theta + \theta^{17} 800K\epsilon)$ and if $\theta = \epsilon^{-18}$, the conclusion follows.

<u>Remark</u>. Lemmas 4 and 6 are valid with the same proof if $b(x, y) = (|x| \wedge |y|)\operatorname{sgn} y$ is substituted for $a(x, y)$ in these lemmas.

It was observed in section 2 that if a Banach space which is an $\mathcal{L}_{p,\lambda}$ space for all $\lambda > 1$, then X is linearly isometrically embeddable into a $L_p(\mu)$ space. Thus it will be assumed below that X is a subspace of $L_p(\mu)$. For $y \in L_p(\mu)$, J_y denotes the band projection onto the band generated by y. Thus, for $x \in L_p(\mu)$, $J_y x = x \operatorname{sgn} y$.

<u>Lemma 8</u>. Let $X = L_p(\mu)$ be an $\mathcal{L}_{p,\lambda}$ space for all $\lambda > 1$. If $1 \le p < \infty$, $p \neq 2$, and $x, y \in X$, then $J_y x$, $|x| \operatorname{sgn} y$, and $(\operatorname{Re} x\, \overline{\operatorname{sgn} y})^+ \operatorname{sgn} y$ are all in X.

<u>Proof</u>. Suppose $\epsilon > 0$. Let E be a finite dimensional subspace of X such that there is an isomorphism T of E onto $\ell_p(k)$ such that $\|T\| = 1$ and $\|T^{-1}\| \le \lambda < 2^{1/p}$, where $(\lambda^p - 1)^{1/p} < \epsilon^{19}$. If $S = T^{-1}$, then by Lemma 6, $\|a(x, y) - S(a(Tx, Ty))\| \le A\epsilon(\|Tx\| + \|Ty\|) \le A\epsilon(\|x\| + \|y\|)$. Since ϵ is arbitrary and $S(a(Tx, Ty)) \in E$, it follows that $a(x, y) \in X$.

Now consider $u_n = a(x, ny)$. Then by the dominated convergence theorem $(\operatorname{Re} x\, \overline{\operatorname{sgn} y})^+ \operatorname{sgn} y = \lim u_n$ is in X. Applying this to -x and $\pm$ ix, it follows that $J_y x = \sum_{k=0}^{3} (\text{Re } i^k x \overline{\text{sgn } y})^+$ is in X. Using the remark following Lemma 7 one obtains that $(|x| \wedge |y|)\text{sgn } y \in X$ and, hence, $|x| \text{ sgn } y \in X$.

Proof of Theorem 12. Since the theorem is trivial for $p = 2$, one can assume that $1 \leq p < \infty$, $p \neq 2$. Let D be a maximal pairwise disjoint subset of X. By Lemma 8, $J_y X \subset X$ for all $y \in D$. Moreover, $X = (\oplus \sum_{y \in D} J_y X)_{\ell_p}$. For the net of finite sums, $J_{y_1} x + \cdots + J_{y_n} x$ converges to $J_D x$ and by the maximality of D, $J_D x = x$.

Thus all that is left is to observe that $x \to \overline{x \text{ sgn } y}$ is an isometry of $J_y X$ onto a closed sublattice of $L_p(\mu)$. This follows from Lemma 8 since if $x \in J_y X$, then $|x \overline{\text{sgn } y}| = |x| \text{sgn } y \cdot \overline{\text{sgn } y}$ and $|x| \text{sgn } y \in J_y X$. Also, $(\text{Re } x \overline{\text{sgn } y})^+ = (\text{Re } \overline{\text{sgn } y})^+ \text{sgn } y \, \overline{\text{sgn } y}$ and $(\text{Re } x \text{ sgn } y)^+ \text{ sgn } y \in J_y X$. Since a closed sublattice of $L_p(\mu)$ is an $L_p(\nu)$ for some ν, the result follows.

In [1] Bernau and Lacey gave a condition of almost disjoint preserving isomorphisms (as in Lemma 5) and an abstract version of dominated convergence to characterize Banach spaces with LUST which are isomorphic to a Banach lattice with order continuous norm.

REFERENCES

1. S. J. Bernau and H. E. Lacey, A local characterization of complex Banach lattices with order continuous norm, Studia Math., T. LVIII (1976), 1-28.
2. D. Dacunha-Castelle and J. Krivine, Applications des ultraoroduits a l'etude des espaces et des algebres de Banach, Studia Math., T. XLI (1972), 315-334.
3. E. Dubinsky, A. Pelczynski, and H. P. Rosenthal, On Banach spaces X for which $\Pi_2(\mathcal{L}_\infty, X) = B(\mathcal{L}_\infty, X)$, Studia Math., T. XLIV (1972), 617-648.
4. T. Figiel, W. B. Johnson, and L. Tzafriri, On Banach lattices and spaces having local unconditional structure with applications to Lorentz sequence spaces, J. Approx. Theory, 13 (1975), 297-312.
5. Y. Gordon and D. R. Lewis, Absolutely summing operators and local unconditional structures, Acta. Math., 133 (1974), 27-48.
6. W. B. Johnson, On finite dimensional subspaces of Banach spaces with local unconditional structure, Studia Math., 51 (1974), 225-240.
7. W. B. Johnson and L. Tzafriri, Some more Banach spaces which do not have local unconditional structure, preprint.
8. J. L. Krivine, Sur les espaces isomorphes a ℓ^p, Séminaire Maurey-Schwartz, 1974-75, No. 12.
9. H. Elton Lacey, The Isometric Theory of Classical Banach Spaces. Springer-Verlag, Grund. Band 208, New York, 1974.
10. J. Lindenstrauss and A. Pelczynski, Absolutely summing operators in ℓ_p-spaces and their applications, Studia Math., T. XXIX (1968), 275-326.
11. J. Lindenstrauss and H. P. Rosenthal, The $\mathcal{L}_p$ spaces, Israel J. Math., 7 (1969), 325-349.
12. B. Maurey and G. Pisier, Caractérisation d'une classe d'espaces de Banach par des propriétés de series aléatoires vectorielles, C. R. Acad. Sci., Paris, t. 277 (1973), Serie A, 687-690.
13. P. Meyer-Nieberg, Charakterisierung einiger topologischer und ordnungs-theoretischer Eigenshaften von Banach-verbänden mit Hilfe disjunkter Folgen., Arch. Math., 24 (1973), 640-647.
14. A. Pelczynski, Banach spaces of analytic functions and absolutely summing operators, Lectures notes of Kent Conference, August, 1976, p. 155.

15. A. Pelczynski, Projections in certain Banach spaces, Studia Math., 19 (1960), 209-228.

16. A. Pietsch, Absolutely p-summing operators in $\mathcal{L}_p$-spaces, Bull. Soc. Math., France Mem., 31-32 (1972), 285-315.

17. J. R. Retherford, Applications of Banach ideals of operators, Bull. Amer. Math. Soc., 81 (1975), 978-1012.

18. Schaefer, H. H., Banach Lattices and Positive Operators. Springer-Verlag, New York, Grund. Band. 215 (1974).

19. J. Stern, Propriétés locales et ultrapuissances d'espaces de Banach. Séminaire Maurey-Schwartz (1974-75), No. 7.

20. L. Tzafriri, Reflexivity in Banach lattices and their subspaces, J. Functional Analysis, 10 (1972), 1-18.

21. ______, Remarks on contractive projections in L_p spaces, Israel J. Math., 7 (1969), 9-15.

22. M. Zippin, On bases in Banach spaces. Thesis: Hebrew University, Jerusalem, Israel.

DUALS OF TENSOR PRODUCTS

D. R. Lewis[1]
The Ohio State University
Columbus, OH 43210/USA

Abstract: A number of tensor norms α for which the duality relation $(E \overset{\alpha}{\otimes} F)' = E' \overset{\alpha'}{\otimes} F')$ holds are described. This duality is used to investigate weak sequential completeness and reflexivity in tensor products.

One of the results of Grothendieck [5] is that

$$(E \overset{\alpha}{\otimes} F)' = E' \overset{\alpha'}{\otimes} F'$$

for α' the greatest tensor norm, whenever F' has both the approximation property and the Radon-Nikodym property. This paper is concerned with describing some other tensor norms α for which this duality is true. Roughly speaking it turns out that if this identification of the dual $E \overset{\alpha}{\otimes} F$ works for $F = c_0$, then a large number of tensor norms can be constructed from α for which the duality relation holds. The case $F = c_0$ is easily checked for many of the naturally occuring tensor norms. The duality results are applied to prove that several Banach ideal norms produce reflexive spacesof operators. For instance if E and F are both reflexive and have a.p., then the space of operators which admit factorizations of the form

$$E \to Q \to L \to F,$$

L on L_1-space and Q a quotient of a L_p-space $(1 < p < \infty)$, is reflexive under the factorization norm. Some results on the weak completeness of $E \overset{\alpha}{\otimes} F$ are also given; for instance, if E is weakly sequentially complete, then so is $L_p(\mu) \hat{\otimes} E$ for $1 < p < \infty$.

All of the basic properties of tensor products used below are implicit in Grothendieck's Resumé [6], although it is often impossible to supply a specific reference. Readable expositions of the basic results of [6] may be found in [1] and the first sections of [14].

The notation and terminology is primarily that of [6]. The symbol α always represents a tensor norm ($\otimes$-norm) and $|\ |_\alpha$ is the associated reasonable norm on $E \overset{\alpha}{\otimes} F$, the α-tensor product of Banach spaces E and F. The greatest and least $\otimes$-norms are $\wedge$ and $\vee$, respectively, and α', ${}^t\alpha$ and $\alpha^\vee = ({}^t\alpha)'$ are the dual, transpose and contragradient $\otimes$-norms of α, respectively. For α a $\otimes$-norm $\alpha\backslash$, $/\alpha$, $\alpha/$ and $\backslash\alpha$ denote, respectively, the right injective, left injective, right projective and left projective hulls of α. The functional on $E' \hat{\otimes} E$ induced by the action $\langle\, ,\, \rangle$ of E' on E is denoted by Tr. The space of α-integral operators from E to F is written $L^\alpha(E,F)$ (or $\alpha(E,F)$) and $\|u\|_\alpha$ (or $\alpha(u)$) is the α-integral norm of u.

[1] Research partially supported by NSF MCS 75-06948 A02.

The space $L^{\alpha}(E, F')$ is identified with the dual of $(E \overset{\alpha'}{\otimes} F)'$ in the following manner. For $u\colon E \to F'$ α-integral the Kronecker product $u \otimes 1\colon E \otimes F \to F' \otimes F$ extends to a continuous map of norm $\leq \|u\|_{\alpha}$ from $E \overset{\alpha'}{\otimes} F$ into $F' \hat{\otimes} F$. The action of u on $v \in E \overset{\alpha'}{\otimes} F$ is taken to be $\langle v, u\rangle = \langle (u \otimes 1)(v), Tr\rangle$.

There is always a canonical map

$$E' \overset{\alpha}{\otimes} F' \longrightarrow L^{\alpha}(E,F')$$

which sends a simple tensor $x' \otimes y'$ to the operator defined by $u(x) = \langle x,x'\rangle y'$. This natural map has norm one and is one-to-one if E' or F' have the approximation property (a.p.) and is an into isometry if both E' and F' have the metric approximation property (m.a.p.). If this natural map is an onto isometry we write

$$E' \overset{\alpha}{\otimes} F' = L^{\alpha}(E, F').$$

<u>Definition</u>: A $\otimes$-norm α has the Radon-Nikodym-Property (r.n.p.) if $E' \overset{\alpha}{\otimes} \ell_1 = L^{\alpha}(E, \ell_1)$ for every space E.

Hopefully the reason for this terminology will be apparent from the statement and proof of Theorem 3. The reason for considering such norms is two-fold. Many of the standard $\otimes$-norms have r.n.p., and once one such norm is known, other $\otimes$-norms with r.n.p. can naturally be constructed from it. Second, it turns out that if α has r.n.p. then the duality

$$E' \overset{\alpha}{\otimes} F' = L^{\alpha}(E, F') = (E \overset{\alpha'}{\otimes} F)'$$

holds for many spaces F other than $F = c_0$.

The basic examples of $\otimes$-norms we consider are the norms i_{pq} and γ_p.

Let $1 \leq p \leq q \leq \infty$. A operator $\varphi\colon E \to F$ is i_{pq}-<u>integral</u> if $\psi\varphi$($\psi\colon F \to F''$ the canonical embedding) admits a factorization

$$E \xrightarrow{w} L_p(\mu) \xrightarrow{v} L_q(\mu) \xrightarrow{u} F''$$

for some probability measure μ, with v the inclusion. The i_{pq}-<u>integral norm</u> is $i_{pq}(\varphi) = \inf \|u\| \, \|w\|$, with the infimum taken over all such factorizations.

Let $1 \leq p \leq \infty$. An operator $\varphi\colon E \to F$ is γ_p-<u>integral</u> if $\psi\varphi$ has a factorization

$$E \xrightarrow{v} L_p(\nu) \xrightarrow{u} F''$$

for some measure ν. The γ_p-<u>integral norm</u> is $\gamma_p(u) = \inf \|v\| \, \|u\|$, with the infimum taken over all such factorizations.

Every dual space F is norm one complemented in its bidual, so in case F is a dual, the operator u occuring in each typical factorization can be taken to map into F rather than F''.

It is known that both γ_p and i_{pq} are maximal Banach ideal norms (Kwapien [7] for γ_p and J. T. Lapreste for i_{pq}) so it is an exercise to show that both norms define $\otimes$-norms. Recall that a $\otimes$-norm α is termed <u>accessible</u> if the natural inclusion of

$E \overset{\alpha}{\otimes} F$ into $(E' \overset{\alpha'}{\otimes} F')'$ is an into isometry whenever E or F is finite dimensional. Using the integral representation formula for $i_{pq}^{\vee}$ and $\gamma_p^{\vee}$ ([4] and [7], it can be shown that $i_{pq}^{\vee}$ and $\gamma_p^{\vee}$ are accessible (see [14] Corollary 5, p. 85, for a proof for $i_{\infty p}$).

Lemma 1. Suppose that either E' has m.a.p. or that α is an accessible $\otimes$-norm. If $u: E \to F$ is α-integral and $v: F \to G$ is in the norm closure of the finite rank operators, then vu is the image, under the canonical inclusion of $E' \overset{\alpha}{\otimes} G$ into $L^{\alpha}(E, G)$, of a $t \in E' \overset{\alpha}{\otimes} G$ satisfying $|t|_{\alpha} \leq \|u\|_{\alpha}\|v\|$.

Proof. By [6] the assumption that E' has m.a.p. or α is accessible shows that if $w: E' \to G'$ is any $(\alpha^t)^{\vee} = \alpha'$-integral operator, then wu' is $\wedge$-integral and $\|wu'\|_{\wedge} \leq \|w\|_{\alpha'}\|u'\|_{\alpha^t}$. Thus defining $\psi: L^{\alpha'}(E', G') \to L^{\wedge}(F', G')$ by $\psi(w) = wu'$ gives an operator with $\|\psi\| \leq \|u'\|_{\alpha^t} = \|u\|_{\alpha}$. Consider ψ as a map from $(E' \overset{\alpha}{\otimes} G)' = L^{\alpha'}(E', G')$ into $(F' \check{\otimes} G)' = L^{\wedge}(F', G')$. Checking simple tensors shows that $\psi'|F' \check{\otimes} G = u' \otimes 1$ maps $F' \check{\otimes} G$ into $E' \overset{\alpha}{\otimes} G$ with norm $\leq \|u\|_{\alpha}$. Now simply identify v with an element of $F' \check{\otimes} G$ and let $t = (u' \otimes 1)(v)$. The commutativity of the diagram

$$\begin{array}{ccc} F' \check{\otimes} G & \xrightarrow{u'\otimes 1} & E' \overset{\alpha}{\otimes} G \\ \downarrow & & \downarrow \\ L(F, G) & \xrightarrow[w \to uw]{} & L^{\alpha}(E,G) \end{array} ,$$

where the unlabeled maps are the natural inclusions, proves the lemma.

Theorem 2. The following $\otimes$-norms have r.n.p.

(1) i_{pq} for $1 < \max(p, q)$.

(2) $i_{pq}^{\vee}$ for $1 \leq p < \infty$.

(3) γ_p for $1 < p \leq \infty$ and $\gamma_p\backslash$ for $1 < p < \infty$.

(4) $\gamma_p^{\vee}$ for $1 \leq p < \infty$.

Proof. In the various parts of the proof α denotes the $\otimes$-norm in question.

(1) $\alpha = i_{pq}$, $1 \leq q \leq p \leq \infty$: Given $\varphi: E \to \ell_1$ i_{pq}-integral and $\lambda > 1$ write $\varphi = uvw$ as in the definition with $\|u\|\,\|w\| \leq \lambda\, i_{pq}(\varphi)$. Let $(e_i) \subset \ell_{\infty}$ be the sequence of unit vectors of ℓ_{∞}. We will show first that $uv = u_2u_1$, where u_2 is a compact operator on ℓ_1, $\|u_2\| \leq 1$ and $i_{pq}(u_1) \leq \lambda\|u\|$.

Case (a), $1 < q$: Here $u: L_q(\mu) \to \ell_1$ is weakly compact and hence compact by the Schur lemma, so

$$\lim_n \sup_{\|x\|\leq 1} \Sigma_{i\geq n} |\langle u(x), e_i\rangle| = 0.$$

From this it follows easily that there is a positive sequence $(a_i) \in c_0$ of norm one which satisfies

$$\sup_{\|x\|\le 1} \Sigma_{i\ge 1}\, a_i^{-1}\, |\langle u(x),\, e_i\rangle| \le \lambda\|u\|.$$

Define v_1: $L_q(\mu) \to \ell_1$ by $v_1(x) = (a_i^{-1}\langle u(x), e_i\rangle)_{i\ge 1}$, let $u_1 = v_1 v$ and define u_2: $\ell_1 \to \ell_1$ by $u_2(z) = (a_i\langle z,\, e_i\rangle)_{i\ge 1}$. The operators u_1 and u_2 have the desired properties.

Case (b), $1 = q \le p \le \infty$. Observe that $\Sigma_{i\ge 1}|u'(e_i)| \le \|u\|$, μ - a.e. . For $k \ge 1$, let

$$a_k^{p'} = \int |u'(e_k)|\; [\Sigma_{i\ge 1}|u'(e_i)|]^{p'-1}\, d\mu,$$

where $1/p + 1/p' = 1$. Since $(a_k) \in \ell_{p'}$ has norm at most $\|u\|$, there is a positive sequence $(b_k) \in c_0$ of norm one so that $(a_k b_k^{-1}) \in \ell_{p'}$ has norm at most $\lambda\|u\|$. Define v_1: $L_p(\mu) \to \ell_p$ by $v_1(f) = (a_k^{-1}\langle f,\, (uv)'(e_k)\rangle)$, v_2: $\ell_p \to \ell_1$ by $v_2((x_i)) = (a_i b_i^{-1} x_i)$ and u_2: $\ell_1 \to \ell_1$ by $u_2((x_i)) = (b_i x_i)$. Clearly u_2 is compact and $\|u_2\| \le 1$. Writing $u_1 = v_2 v_1$, $uv = u_2 u_1$ and $i_{p1}(u_1) \le \|v_1\| i_{p1}(v_2) \le \|v_1\|\lambda\|u\|$, so it only remains to show that $\|v_1\| = \|v_1'\| \le 1$. But for $(x_i) \in \ell_p'$ a norm one element,

$$\|v_1'((x_i))\|^{p'} \le \int |\Sigma_{i\ge 1}\, a_i^{-1}|x_i|\; |u'(e_i)|^{1/p'}|u'(e_i)|^{1/p}|^{p'}\, d\mu$$

$$\le \int (\Sigma_{k\ge 1} a_k^{-p'}|x_k|^{p'}|u'(e_k)|)(\Sigma_{i\ge 1}|u'(e_i)|)^{p'/p}\, d\mu$$

$$= \Sigma_{k\ge 1}|x_k|^{p'}\; a_k^{-p'} \int |u'(e_k)|[\Sigma_{i\ge 1}|u'(e_i)|]^{p'-1}\, d\mu = 1.$$

Now consider the commutative diagram

$$\begin{array}{ccccc} & L_p(\mu)' \overset{\alpha}{\otimes} \ell_1 & \longrightarrow & L^{\alpha}(L_p(\mu),\, \ell_1) & \\ w' \otimes 1 & \downarrow & & \downarrow & a \to aw \\ & E' \overset{\alpha}{\otimes} \ell_1 & \longrightarrow & L^{\alpha}(E,\, \ell_1), & \end{array}$$

where the unlabeled arrows are the canonical inclusions. Since $L_p(\mu)'$ has m.a.p. Lemma 1 shows that $uv = u_2 u_1 \in L^{\alpha}(L_p(\mu),\, \ell_1)$ is the image of a $t \in L_p(\mu)' \overset{\alpha}{\otimes} \ell_1$ for which $|t|_{\alpha} \le \|u_2\|\|u_1\|_{\alpha} \le \lambda\; \|u\|$. Then $(w' \otimes 1)(t)$ goes to $(uv)w = \varphi$ under the natural map and $|(w' \otimes 1)(t)|_{\alpha} \le \lambda\|w'\|\|u\| \le \lambda^2\; i_{pq}(\varphi)$. Since $\lambda > 1$ was arbitrary and the inclusion $E' \overset{\alpha}{\otimes} \ell_1 \hookrightarrow L^{\alpha}(E,\, \ell_1)$ is one-to-one (ℓ_1 has a.p.), the canonical inclusion is an onto isometry. This proves (1).

(2) $\alpha = \ell_{pq}^{\vee}$, $1 \le p < \infty$: From [4], phrased in terms of tensor products, the $i_{pq}^{\vee}$-integral operators φ into ℓ_1 are exactly those which admit a factorization $\varphi = uv$ with v $i_{\infty q'}^{\vee}$-integral and u $i_{p'1}$-integral, and $i_{pq}^{\vee}(\varphi) = \inf i_{\infty q'}^{\vee}(v)\; i_{p'1}(u)$. The argument in case (b) above shows that $u = u_2 u_1$ with u_2 a compact operator on

ℓ_1 and $\|u_2\| i_{p'1}(u_1) \leq \lambda\, i_{p'1}(u)$, $\lambda > 1$ arbitrary. Then $\varphi = u_2(u\, v)$ with $\|u_2\| i^{\vee}_{pq}(u_1 v) \leq \lambda\, i^{\vee}_{pq}(\varphi)$ and Lemma 1 may be applied again because $i^{\vee}_{pq}$ is accessible.

(3) and (4): The proof for γ_p, $1 < p \leq \infty$, is exactly the same as (1) in case (a). The $\gamma_p\backslash$-integral operators are those which factor as $\varphi = uv$, v into a subspace of an $L_p(\mu)$-space, so the same proof applies as well to $\gamma_p\backslash$ for $1 < p < \infty$. The proof for $\gamma^{\vee}_p$ is the same as (2), using Kwapien's factorization theorem [7] for the $\gamma^{\vee}_p$-integral operators.

A Banach space G has the <u>Radon-Nikodym Property</u> (r.n.p.) if every operator u: $L_1(\mu) \to G$, μ any finite measure, has an integral representation $u(f) = \mu(fh)$ for h a μ-Bochner integrable function into G. Examples include separable duals, reflexive spaces and the spaces $\ell_1(\Gamma)$. (cf. [3] and its references).

<u>Theorem 3</u>. Let α be a $\otimes$-norm with r.n.p. and let β be a $\otimes$-norm obtained from α by applying some finite sequence of the operations $\gamma \to \gamma/$, $\gamma \to /\gamma$ and $\gamma \to \backslash\gamma$. Then for any space E,

$$E' \overset{\beta}{\otimes} F' = L^{\beta}(E, F') = (E \overset{\beta'}{\otimes} F)'$$

if F' has r.n.p. and a.p.

The proof of the theorem (parts (c), (d) and (e) below) will also prove

<u>Theorem 4</u>. If α is a $\otimes$-norm with r.n.p. then $\alpha/$, $\backslash\alpha$ and $/\alpha$ also have r.n.p.

It is convenient to divide the proof of Theorem 3 into a number of steps.

(a) The conclusion holds for $\alpha = \beta$ and $F = c_0(\Gamma)$, Γ any set.

<u>Proof</u>. An α-integral operator u: $E \to F' = \ell_1(\Gamma)$ is compact. In fact, if $(x_i) \subset E$ is a bounded sequence then $(u(x_i))$ is contained in a separable subspace of $\ell_1(\Gamma)$ which is isometric to ℓ_1 and which is the image of a norm one projection v. The composition vu is an α-integral operator into ℓ_1 and, since α has r.n.p., vu is α-nuclear and hence compact. But then u is compact because $vu(x_i) = u(x_i)$ for each i. In particular u(E) is separable, so again there is a norm one projection w of $\ell_1(\Gamma)$ onto a subspace G which contains u(E) and is isometric to ℓ_1. The astriction u_a of u to an operator from E into G satisfies $\|u_a\|_\alpha = \|u\|_\alpha$ since G is norm one complemented. Statement (a) now follows from the commutativity of the diagram

$$\begin{array}{ccc} E' \overset{\alpha}{\otimes} \ell_1(\Gamma) & \longrightarrow & L^{\alpha}(E, \ell_1(\Gamma)) \\ \uparrow & & \uparrow \\ E' \overset{\alpha}{\otimes} G & \longrightarrow & L^{\alpha}(E, G) \end{array}$$

of natural maps, because u_a is the image of a $t \in E' \overset{\alpha}{\otimes} G$ with $|t|_{\alpha} = \|u_a\|_{\alpha} = \|u\|_{\alpha}$.

(b) The conclusion of the theorem holds for $\beta = \alpha/$.

Proof. Given u: $E \to F'$ $\alpha/$-integral and any $\lambda > 1$, u has a factorization

$$E \xrightarrow{v} L_1(\mu) \xrightarrow{w} F'$$

which satisfies $\|v\|_{\alpha}\|w\| \leq \lambda\|u\|_{\alpha/}$ ([6], Corollary 1, p. 32). The space F' has r.n.p. so it follows from [8] (see [12] for a nice proof) that there is a set Γ and operators s: $L_1(\mu) \to \ell_1(\Gamma)$, t: $\ell_1(\Gamma) \to F'$ such that $\|s\|\|t\| \leq \lambda\|w\|$ and $w = ts$. Consider the commutative diagram

$$\begin{array}{ccccc} & E' \overset{\alpha}{\otimes} \ell_1(\Gamma) & \longrightarrow & L^{\alpha}(E, \ell_1(\Gamma)) & \\ 1 \otimes t & \downarrow & & \downarrow & a \to ta \\ & E' \overset{\alpha/}{\otimes} F' & \longrightarrow & L^{\alpha/}(E, F'), & \end{array}$$

where the unmarked arrows are the canonical inclusions. The composition sv is α-integral and $\|sv\|_{\alpha} \leq \|s\|\|v\|_{\alpha}$, so by (a) sv is the image of a $b \in E' \overset{\alpha}{\otimes} \ell_1(\Gamma)$ with $|b|_{\alpha} \leq \|s\|\|v\|_{\alpha}$. But then $u = t(sv)$ is the image of $(1 \otimes t)(b)$ and $|(1 \otimes t)(b)|_{\alpha} \leq \lambda^2\|u\|_{\alpha/}$. Since $\lambda > 1$ is arbitrary and the canonical inclusion $E' \overset{\alpha/}{\otimes} F' \hookrightarrow L^{\alpha/}(E, F')$ is one-to-one (F' has a.p.), this shows that the canonical inclusion is on onto isometry.

(c) If the conclusion holds for β then it holds for $\beta/$.

Proof. Taking $F = c_0$ shows that β has r.n.p. so this follows from (b).

(d) If the conclusion holds for β then it holds for $\backslash\beta$.

Proof. For u: $E \to F'$ $\backslash\beta$-integral and $\lambda > 1$, u has a factorization

$$E \xrightarrow{v} C(S) \xrightarrow{w} F'$$

with $\|v\|\|w\|_{\beta} \leq \lambda\|w\|_{\backslash\beta}$ ([6]). The diagram of natural maps

$$\begin{array}{ccccc} & C(S)' \overset{\beta}{\otimes} F' & \longrightarrow & L^{\beta}(C(S), F') & \\ v' \otimes 1 & \downarrow & & \downarrow & a \to av \\ & E' \overset{\backslash\beta}{\otimes} F' & \longrightarrow & L^{\backslash\beta}(C(S), F') & \end{array}$$

is commutative and the top horizontal arrow is an onto isometry, so arguing as in (b) establishes (d).

(e) If the conclusion is true for β then it is true for $/\beta$.

Proof. Let u: $E \to F'$ be $/\alpha$-integral, L be an $L_1(\mu)$-space for which there is a quotient map φ: $L \to E$. Consider the diagram

$$\begin{array}{ccc} E' \overset{/\alpha}{\otimes} F' & \longrightarrow & L^{/\alpha}(E, F') \\ \varphi' \otimes 1 \downarrow & & \downarrow \quad a \to a\varphi \\ L' \overset{\alpha}{\otimes} F' & \longrightarrow & L^{\alpha}(L, F'), \end{array}$$

where the unmarked are the canonical inclusions. It is known [6] that both vertical arrows are into isometrics. Since $\|u\varphi\|_\alpha = \|u\|_{/\alpha}$ the hypothesis about β guarantees that $u\varphi$ is the image of a $t \in L' \overset{\alpha}{\otimes} F'$ with $|t_\alpha| = \|u\|_{/\alpha}$. To prove (e), it is enough to show that t lies in the image of $E' \overset{/\alpha}{\otimes} F'$ under the isometry $\varphi' \otimes 1$. To this end let $v \in (L' \overset{\alpha}{\otimes} F')'$ be a functional vanishing on the range of $\varphi' \otimes 1$. Consider v as an element of $L^{\alpha'}(L', F'')$ so that $v\varphi' = 0$. Let $v \otimes 1$ denote the operator from $L' \overset{\alpha}{\otimes} F'$ into $F'' \hat{\otimes} F'$ which satisfies $(v \otimes 1)'(Tr) = v$, and write $a \to a^{\sim}$ for the natural map of $F'' \hat{\otimes} F'$ into $L(F')$. Write $a = (v \otimes 1)(t)$, so that $(a^{\sim})' = (v\varphi')u = 0$. Since F' has a.p. this implies ([5], p. 165) that $\langle a, Tr \rangle = 0$, and so $\langle t, v \rangle = \langle a, Tr \rangle = 0$. Since v was arbitrary the Hahn-Banach theorem shows that $t = (\varphi' \otimes 1)(s)$ for some $s \in E' \overset{/\alpha}{\otimes} F'$, and $|s|_{/\alpha} = |t|_\alpha = \|u\|_{/\alpha}$ since $\varphi \otimes 1$ is an isometry.

Corollary 5. Let α be a $\otimes$-norm such that both α and $\alpha^{\vee}$ have r.n.p. Let β be either $(/\alpha)/$ or $/(\alpha/)$, or the dual, transpose or contragradient of either norm. Then $L^{\beta}(E,F)$ and $E \overset{\beta}{\otimes} F$ are reflexive if E and F are reflexive and have a.p.

Proof. Let $\beta = (/\alpha)/$. By Theorem 4, $/\alpha$ has r.n.p. so

$$L^{\beta}(E,F) = E' \overset{\beta}{\otimes} F = (E \overset{\beta'}{\otimes} F')',$$

by Theorem 3. Since $\beta^{\vee} = /(\alpha^{\vee}/)$ and $\alpha^{\vee}/$ has r.n.p. (Theorem 4), Theorem 3 also yields

$$L^{\beta^{\vee}}(F,E) = F' \overset{\beta^{\vee}}{\otimes} E = (F \overset{t_\beta}{\otimes} E')'.$$

Combining equalities (the natural isometries)

$$(E \overset{\beta'}{\otimes} F')'' = (E' \overset{\beta}{\otimes} F)' = (F \overset{t_\beta}{\otimes} E')' = F' \overset{\beta^{\vee}}{\otimes} E = E \overset{\beta'}{\otimes} F'$$

so that the β' (and hence β, ${}^t\beta$ and $\beta^{\vee}$) tensor product of reflexive spaces is reflexive. But then $L^{\beta}(E, F) = (E \overset{\beta'}{\otimes} F')'$ is reflexive, and similarly for β, ${}^t\beta$ and $\beta^{\vee}$. The case $\beta = /(\alpha/)$ follows by interchanging the roles of α and $\alpha^{\vee}$, since $\alpha^{\vee\vee} = \alpha$.

Corollary 6. Let E and F be reflexive spaces which both have a.p. For $1 < p < \infty$ and $\beta = (/\gamma_p)/$, $(/\gamma_p\backslash)/$, $\backslash(\gamma_p\backslash)$ or $\backslash(/\gamma_p\backslash)$, $L^{\beta}(E,F)$ is reflexive.

Proof. For $1 < p < \infty$, γ_p and $\gamma_p^{\vee}$ have r.n.p. by Theorem 2. Since $(\gamma_p\backslash)^{\vee} = \backslash(\gamma_p^{\vee})$ both $\gamma_p\backslash$ and its contragradient have r.n.p. by Theorems 2 and 4. Thus the preceeding corollary applies with $\beta = (/\gamma_p)/$ and $(/\gamma_p\backslash)/$. But since ${}^t[(/\gamma_p)/] = (\gamma_q\backslash)$ and ${}^t[(/\gamma_p\backslash)/] = \backslash(/\gamma_q\backslash)$, $1/p + 1/q = 1$, the corollary applies to the other two $\otimes$-norms as well.

The $(/\gamma_p)/$-integral operators, etc., can easily be described in terms of a typical factorization. An operator φ: $E \to F$ is $\gamma_p/$-integral (respectively, $(/\gamma_p)/$, $(\gamma_p\backslash)/$, $(/\gamma_p\backslash)/$-integral) if $\psi\varphi$(ψ: $F \to F''$ the natural embedding) has a factorization

$$E \xrightarrow{u} A \xrightarrow{v} L_1(\mu) \xrightarrow{w} F''$$

for some measure μ and some space A which is isometric to an $L_p(\nu)$-space (resp., a quotient of, a subspace of, a subspace of a quotient of, an $L_p(\nu)$-space). The norm of φ is the infimum, taken over all such factorizations, of $\|u\|\|v\|\|w\|$. Similarly, a $\backslash\gamma_p$-integral (resp., $\backslash(/\gamma_p)$, $\backslash(\gamma_p\backslash)$, $\backslash(/\gamma_p\backslash)$-integral) operator is one which has a factorization

$$E \xrightarrow{u} L_\infty(\mu) \xrightarrow{v} B \xrightarrow{w} F''$$

for some measure μ and space B which is isometric to an $L_p(\nu)$-space (resp., a quotient of, a subspace of, a subspace of a quotient of, an $L_p(\nu)$-space). In case $p = 2$, the descriptions are much simplier since A and B may be taken to be Hilbert spaces. More formally, since $\gamma_2 = \gamma_2\backslash = /\gamma_2 = /\gamma_2\backslash$,

<u>Corollary 7</u>. If E and F are reflexive and have a.p., then $(\gamma_2/)(E, F)$ and $(\backslash\gamma_2)(E,F)$ are reflexive.

Corollary 5 may also be applied with the $\otimes$-norm $\alpha = i_{pq}$ in case $1 \leq q \leq p < \infty$ and $1 < \max(p, q)$. The most interesting case is $q = 1$, which shows that the $\otimes$-norms i_{pq} and $/i_{p1}$ yield reflexive $\otimes$-products for $1 < p < \infty$. But these results also follow directly from the theorem of Persson [9] since ${}^t i_{pq} = p'$-integral norm and ${}^t(/i_{p1}) = p'$-absolutely summing norm.

<u>Theorem 8</u>. Let α and β be related as in Theorem 3. If E and F are both weakly sequentially complete (w.s.c.) and if F has an unconditional basis, then $E \overset{\beta}{\otimes} F$ is w.s.c. .

<u>Proof</u>. By a theorem of James (cf. [2]) $F = G'$ for some space G. In particular, F has r.n.p. and a.p., so by Theorem 3

$$E \overset{\beta}{\otimes} F \subset E'' \overset{\beta}{\otimes} G' = L^\beta(E', G') = (E' \overset{\beta'}{\otimes} G)',$$

with all identifications natural and isometric. Let (u_k) be a weakly Cauchy sequence in $E \overset{\beta}{\otimes} F$. The sequence is norm bounded when considered in $(E' \overset{\beta'}{\otimes} G)'$ and thus has a weak-star cluster point $u \in E'' \overset{\beta}{\otimes} F$. We show that $u \in E \overset{\beta}{\otimes} F$ and that

(*) $$\langle u_k, x' \otimes y' \rangle \to \langle u, x' \otimes y' \rangle \text{ for all } x' \in E' \text{ and } y' \in F'.$$

To see this consider u and each u_k as elements of $L^\beta(E', G')$. For $x' \in E'$ $(u_k(x'))$ is weakly Cauchy since (u_k) is weakly Cauchy. Then $(u_k(x'))$ converges weakly in $F = G'$ and clusters weak-star to $u(x')$, so $u_k(x') \to u(x')$ weakly in F. This proves (*) which implies $u_k'(y') \to u'(y')$ weak-star in E'' for each $y' \in F'$.

But since E is w.s.c. and $(u_k'(y')) \subset E$ is weakly Cauchy, $u'(y')$ is actually in E. Now since $u'(F') \subset E$ and F has a.p., the same argument used in (e) of the proof of Theorem 3 shows that $u \in E \overset{\beta}{\otimes} F$.

To show that $u_k \to u$ weakly let $w \in (E \overset{\beta}{\otimes} F)' = L^{\beta'}(E, F')$. Extend $w \otimes 1$ to a continuous operator from $E \overset{\beta}{\otimes} F$ into $F' \hat{\otimes} F$ satisfying $(w \otimes 1)'(Tr) = w$. Let $(e_i) \subset F$ be an unconditional basis with coefficient functionals (e_i'). The unconditionality of the basis shows that the bilinear form $\varphi(y', y) = (\langle e_i, y'\rangle\langle y, e_i'\rangle)_{i\geq 1}$ is continuous from $F' \times F$ into ℓ_1, and thus extends to a continuous operator φ: $F' \hat{\otimes} F \to \ell_1$ satisfying $\varphi'(\psi) = Tr$, where $\psi \in \ell_\infty$ is the constantly one sequence. The sequence $(\varphi(w \otimes 1)(u_k))_{k\geq 1} \subset \ell_1$ is weakly Cauchy and so, by the Schur lemma, converges in norm to some $f \in \ell_1$. By (*), $(\varphi(w \otimes 1)(u_k))_{k\geq 1}$ converges coordinatewise to $\varphi(w \otimes 1)(u)$. This implies that $f = \varphi(w \otimes 1)(u)$ and that $\langle u_k, w\rangle = \langle\varphi(w \otimes 1)(u_k), \psi\rangle$ converges to $\langle\varphi(w \otimes 1)(u), \psi\rangle = \langle u, w\rangle$.

<u>Corollary 9</u>. Let α and β be related as in Theorem 3. If E' and F are w.s.c. and F has an unconditional basis, then $L^{\beta}(E, F)$ is w.s.c. .

<u>Proof</u>. In this case $L^{\beta}(E, F) = E' \overset{\beta}{\otimes} F$, by Theorem 3.

<u>Remarks</u>. (1) Under the hypothesis of Corollary 9, $L^{\beta}(E, F)$ is reflexive iff it has no subspace isomorphic to ℓ_1. This follows from the deep result of Rosenthal [13] that if $\ell_1 \not\subset G$, then every sequence in G has a weakly Cauchy subsequence.

(2) Theorem 8 and its corollary are valid if $F = L_p(\mu)$ for some $1 < p < \infty$ and measure μ. In fact, given a sequence $(u_i) \subset L^{\beta}(E, F)$, there is a separable, norm one complemented sublattice $G \subset F$ which contains each $u_i(E)$. This means that (u_i) can be considered as a sequence in $L^{\beta}(E, G)$, and G has an unconditional basis since it is a separable $L_p(\nu)$-space.

(3) From Theorems 2, 3, and 4, Theorem 8 and its corollary are true for the $\otimes$-norms $\gamma_p/$ and $(/\gamma_p)/$, $1 < p \leq \infty$; $(\gamma_p\backslash)/$ and $(/\gamma_p\backslash)/$, $1 < p < \infty$; and $i_{pq}/$, $/(i_{pq}/)$ and $(/i_{pq})/$ if $1 < \max(p, q)$. In case $q = 1$, the $i_{p1}/$-integral operators are exactly the adjoints of the p'-integral operators [4], and the $/(i_{p1}/)$-integral operators the adjoints of the p'-absolutely summing operators. Taking $q = 1$ and $p = \infty$, $i_{pq}/ = \wedge =$ greatest $\otimes$-norm. Thus

<u>Corollary 10</u>. If E and F are w.s.c. and F has an unconditional basis, then $E \hat{\otimes} F$ is w.s.c. .

<u>Corollary 11</u>. If E is w.s.c. and $1 < p < \infty$, then $L_p(\mu) \hat{\otimes} E$ is w.s.c. for any measure μ.

References

1. I. Amemiga and K. Shiga, On tensor products of Banach spaces, Kotai Math. Sem. Rep., 9 (1957), 161-178.
2. N. M. Day, Normed Linear Spaces, 3rd edition, Ergeb. der. Math. 21, Springer, New York, 1973.
3. J. Diestel and J. J. Uhl, The Radon-Nikodym theorem for Banach space valued measures, Rocky Mountain J. Math., 6 (1976), 1-46.
4. Y. Gordon, D. R. Lewis, and J. R. Retherford, Banach ideals of operators with applications, J. Func. Anal. 14 (1973), 85-129.
5. A. Grothendieck, Produits tensoriels topologiques et espaces nucléaires, Mem. Amer. Math. Soc., 16 (1955).
6. ______, Résumé de la theorie métrique des produits topologiques, Buletin Soc. Math. Sao Paúlo, 8 (1956), 1-79.
7. S. Kwapien, On operators factorizable through L_p-space, Bull. Soc. Math. France Mémorie, 31-32,(1972), 215-225.
8. D. R. Lewis and C. Stegall, Banach spaces whose duals are isomorphic to $\ell_1(\Gamma)$, J. Func. Anal., 19 (1973), 177-187.
9. A. Persson, On some properties of p-nuclear and p-integral operators, Studia Math., 35 (1969), 214-224.
10. A. Pietsch, Absolut-p-summierende Abbildungen in normierten Räumen, Studia Math., 28 (1967), 333-353.
11. A. Pietsch and A. Persson, p-nukleare and p-integrale Abbildungen in Banachräumen, Studia Math., 33 (1969), 19-62.
12. H. P. Rosenthal, The Banach spaces C(K) and $L^p(\mu)$, Bull. Amer. Math. Soc., 81 (1975), 763-781.
13. ______, A characterization of Banach spaces containing ℓ^1, Proc. Nat. Acad. Sci. U.S.A., 71 (1974), 2411-2413.
14. P. Saphar, Produits tensoriels d'espaces de Banach et classes d'applications linéaires, Studia Math., 38 (1970), 71-100.

CLOSED IDEALS IN RINGS OF ANALYTIC FUNCTIONS SATISFYING A LIPSCHITZ CONDITION

A. L. Matheson
Purdue University Calumet Campus
Hammond, IN 46323/USA

1. Let λ_α, $0 < \alpha < 1$, denote the class of functions analytic in the open disk, continuous in the closed disk and satisfying the following Lipschitz condition:

$$\lim \frac{|f(e^{it_1}) - f(e^{it_2})|}{|t_1 - t_2|^\alpha} = 0, \tag{1}$$

uniformly as $|t_1 - t_2| \to 0$. Under the usual Lipschitz norm, λ_α becomes a Banach algebra. The following theorem of Hardy and Littlewood characterizes the classes λ_α and leads to a norm equivalent to the usual Lipschitz norm and more convenient to work with.

Theorem 1. Let $f(z)$ be a function analytic in the open unit disk. Then $f(z)$ is continuous in the closed disk and $f \in \lambda_\alpha$ if and only if

$$\lim (1 - |z|)^{1-\alpha} |f'(z)| = 0 \tag{2}$$

$$\text{uniformly as } |z| \to 0 .$$

The proof of this theorem can be found in Zygmund [10] or Duren [3]. Defining

$$\|f\|_\alpha = \|f\|_\infty + \sup \{(1 - |z|)^{1-\alpha} |f'(z)| : \ |z| < 1\} \tag{3}$$

yields a Banach algebra norm on λ_α, and the proof of the above theorem guarantees that it is equivalent to the usual Lipschitz norm.

The purpose of this paper is to describe the structure of the closed ideals in the Banach algebras λ_α. Let I be a closed ideal and define $E_I = \{z: \ f(z) = 0$ for all f in I and $|z| = 1\}$. It is easily seen that E_I is a closed set of measure zero which satisfies

$$\int_0^{2\pi} \log (\text{dist } (e^{i\theta}, E_I))d\theta > -\infty. \tag{4}$$

Such sets will be called Carleson sets. Following Rudin [7], in the case of the disk algebra, let G_I be the greatest common divisor of the inner factors of the functions in I. Then the closed ideal I is completely determined by the set E_I and the inner function G_I in the following sense. Let E be the Carleson set and G an inner function. Define

$$I(E, G) = \{f \in \lambda_\alpha: \ f(z) = 0 \text{ for all } z \in E \text{ and } fG^{-1} \in H^\infty\}. \tag{5}$$

It is clear that $I(E, G)$ is a closed (possibly trivial) ideal in λ_α, and that $I \subseteq I(E_I, G_I)$.

it should be mentioned, in view of the following theorem of Havin [4] and Shamoyan [8], that the condition $f\,G^{-1} \in H^\infty$ is stronger than it appears.

Theorem 2. If $f \in \lambda_\alpha$ and G is an inner function such that $f\,G^{-1} \in H^\infty$, then $f\,G^{-1} \in \lambda_\alpha$ and

(6) $$\|f\,G^{-1}\|_\alpha \leq C_\alpha \|f\|_\alpha.$$

The main result is:

Theorem 3. $I = I\,(E_I, G_I)$.

To prove Theorem 3, it will suffice, by the Hahn-Banach theorem, to show that any bounded linear functional T which annihilates I also annihilates some dense subset J of $I\,(E_I, G_I)$.

We consider first the special case of Theorem 3 in which $G_I = 1$. The general result will be postponed to section 4. Section 2 is devoted to describing the relevant properties of the annihilator I. One result of this will be to indicate a relevant choice for the subset J of $I(E_I, 1)$. Section 3 is devoted to proving that J is dense in $I(E_I, 1)$.

2. Let T be any bounded linear functional of norm 1 on λ_α, and for fixed z, $|z| > 1$ define

(7) $$T^+(z) = T\left(\frac{z}{z - w}\right).$$

It is clear that the function $\frac{z}{z - w}$ of w is in λ_α, being in fact analytic on the closed unit disk. A calculation yields $\|\frac{z}{z - w}\|_\alpha < C(|z| - 1)^{-1-\alpha}$ from which follows the estimate

(8) $$|T^+(z)| < C\,(|z| - 1)^{-1-\alpha}.$$

Of course T^+ is analytic for $|z| > 1$.

The idea of considering the function T^+ dates back to Carleman [2], although in the present context, this was first used by Taylor and Williams [9] and then by Korenblyum [5]. One of the features of the function $T^+(z)$ that will be needed is contained in the next lemma.

Lemma 1. $$T(f) = \lim_{s \to 1^+} \frac{1}{2\pi} \int_0^{2\pi} T^+(s\,e^{i\theta}) \quad (e^{i\theta})\,d\theta.$$

The next theorem is essentially Theorem 1.1 of Korenblyum [5], applied to λ_α. Since the proofs of the lemma above and Theorem 4 are only minor modifications of Korenblyum's proofs, they will be omitted.

Theorem 4. Let f be a non-zero function in λ_α and let T be a bounded linear functional on λ_α such that $T(f_g) = 0$ for all g in λ_α. For fixed z, $|z| < 1$, define

(9) $$h(z) = T\left(\frac{z(f(w) - f(z))}{w - z}\right)$$

and $$T^-(z) = \frac{h(z)}{g(z)}.$$

Then

(a) T^+ and T^- are analytic continuations of one another across any arc of the unit circle upon which f does not vanish

(b) $|T^+(z)| \leq C(|z| - 1)^{-1-\alpha}$

(c) $|h(z)| \leq C(1 - |z|)^{-1-\alpha}$.

As corollaries, we have the following:

Corollary 1. Let I be a non-zero ideal in λ_α and let T be an annihilator of I. Then

(a) $T^-(z)$ does not depend on $f \in I$.

(b) The function $T(z)$ defined by

$$T(z) = T^+(z) \quad |z| > 1$$
$$T(z) = T^-(z) \quad |z| < 1$$

is analytic in the Riemann sphere off the zero set of I.

Corollary 2. If the zero set of I is empty, then $I = \lambda_\alpha$.

Corollary 2 follows immediately from Liouville's theorem. We also obtain, following Korenblyum's analysis:

Theorem 5. Let I be an ideal and T an annihilator of I. Then, if $G_I = 1$,

(10) $$|T^+(z)| < C\,[\text{dist}\,(z, E_I)]^{-2(1+\alpha)}.$$

Now let $J = \{f \in \lambda_\alpha : \ |f(z)| < C\,[\text{dist}\,(z, E_I)]^4\}$.

By Theorem 5 and the definition of J, if T annihilates I, then $T^+(re^{i\theta})g(e^{i\theta})$ is uniformly bounded $1 < r < 2$, so, by lemma 1,

(11) $$T(g) = \lim_{r\to 1^+} \frac{1}{2\pi}\int_0^{2\pi} T(re^{i\theta})g(e^{i\theta})f(e^{i\theta})\,d\theta$$
$$= \frac{1}{2\pi}\int_0^{2\pi} T(e^{i\theta})g(e^{i\theta})\,d\theta.$$

Now by Beurling's invariant subspace theorem [1], since $G_I = 1$, the ideal I is dense in H^2. Hence there is a sequence (f_n) in I such that $\|1 - f_n\|_2 \to 0$. Thus

$$T(g) = \frac{1}{2\pi}\int_0^{2\pi} T(e^{i\theta})g(e^{i\theta})\,d\theta$$
$$= \lim_{n\to\infty} \frac{1}{2\pi}\int_0^{2\pi} T(e^{i\theta})g(e^{i\theta})f_n(e^{i\theta})\,d\theta = 0.$$

It follows that if I is a closed ideal in λ_α such that $G_I = 1$, the set J defined above is contained in I. By the remark above, to prove Theorem 3 when $G_I = 1$ it suffices to show that J is dense in $I(E_I, 1)$. An outline of this will be given in section 3. The details will be found in Matheson [6].

3. That the set J is dense in $I(E_I, 1)$ will be a consequence of the following theorem.

Theorem 6. Let $f \in \lambda_\alpha$ and suppose E is a closed set on the unit circle such that $f(z) = 0$ for all $z \in E$. Let $p > 0$ be given. Then for any $\epsilon > 0$ there is a function $f_\epsilon \in \lambda_\alpha$ such that

(a) the inner factors of f and f_ϵ coincide,

(b) $|f_\epsilon(z)| \leq C\,[\text{dist}\,(z, E)]^p$ as $\text{dist}\,(z, E) \to 0$

(c) $\|f - f_\epsilon\| < \epsilon$.

We indicate how Theorem 6 follows from a series of lemmas, although many proofs will be omitted through lack of space. The first lemma is a simple consequence of Theorems 1 and 2.

Lemma 2. Let $f \in \lambda_\alpha$ and let $f = GF$ where G is an inner function and F is an outer function. Then for any $\epsilon > 0$ the functions $F^{1+\epsilon}G$ are in λ_α and

$$\lim_{\epsilon \to 0} \|F^{1+\epsilon}G - F\,G\|_\alpha = 0. \tag{12}$$

By Lemma 2 it suffices to consider functions of the form $f = F^pG$ where F is outer, G is inner, $p > 1$ and $F\,G \in \lambda_\alpha$. The main difficulty in the proof of Theorem 6 is isolated in the lext lemma, which is of independent interest.

Lemma 3. Let $f \in \lambda_\alpha$ be of the form $f = F^pG$ where F is outer, G is inner, $p > 1$, and $F\,G \in \lambda_\alpha$, so that

$$|f(z)| < C\,[\text{dist}\,(z, E)]^\beta \tag{13}$$

for some $\beta > \alpha$ and some constant $C > 0$. Let Γ be any open subset of the unit circle such that the endpoints of each component of Γ lie in E. Define

$$F_\Gamma(z) = \exp\,\{\frac{1}{2\pi}\int_\Gamma \frac{e^{i\theta} + z}{e^{i\theta} - z}\log|F\,(e^{i\theta})|\,d\theta\}. \tag{14}$$

Then there is a positive number N_0 independent of Γ such that if $N > N_0$, the function $f\,F_\Gamma^N$ belongs to λ_α and satisfies

$$|(f\,F_\Gamma^N)'(z)| = \sigma\,((1 - |z|)^{\alpha-1}) \tag{15}$$

as $|z| \to 1^-$ uniformly with respect to Γ.

We also need the next lemma

Lemma 4. Let $f \in \lambda_\alpha$, and let E_0 be a finite set in the unit circle on which f vanishes. Let $N > 0$ be given. Then for each $\epsilon > 0$ there is an outer function Φ in

λ_α such that

(a) $\|\Phi f - f\|_\alpha < \epsilon$

(b) $\{z: \Phi(z) = 0\} = E_0$

(c) $|\Phi(z)| < C\,[\text{dist}\,(z, E_0)]^N$ as dist $(z, E_0) \to 0$.

To prove Lemma 4, assume first that $E = \{1\}$ and $N = 1$. Then a calculation yields that $\Phi(z) = \frac{z-1}{z-1-\delta}$ works, where $\delta = \delta(\epsilon)_0$. The lemma then follows by induction.

We now proceed to the proof of Theorem 6. By Lemma 2 it suffices to prove the theorem for $f = F^pG$, when $F\,G \in \lambda_\alpha$. Let (I_k) be the sequence of complementary intervals to the set E. For each $n > 0$, let $B_n = \bigcup_{k=n+1}^{\infty} I_k$ and define

$$F_n(z) = \exp\left\{\frac{N}{2\pi}\int_{B_n} \frac{e^{i\theta}+z}{e^{i\theta}-z}\log|F(e^{i\theta})|\,d\theta\right\}, \tag{16}$$

where $N \geq \max\{N_0, P\,\alpha^{-1}\}$. It follows from Lemma 3 that $f\,F_n \in \lambda_\alpha$ and it is a simple matter, using the estimate (15) and the fact that the functions F_n converge uniformly to 1 on compact subsets of the open disk, to show that $\lim_{n\to\infty} \|f - f\,F_n\|_\alpha = 0$.

The functions $f\,F_n$ clearly satisfy (a) and (c) of Theorem 6 if n is sufficiently large, but not necessarily (b). To rectify this we use Lemma 4, together with the following remark. Let $E_1 = E - \overline{B}_n$ and $E_2 = E \cap \partial\,\overline{B}_n$. Then $E_1 \cup E_2$ is a finite set upon which the function $f\,F_n$ vanishes. Thus given $\epsilon > 0$ there is an outer function Φ in λ_α such that $\|f\,F_n - f\,F_n\Phi\|_\alpha < \epsilon$, and

$$|\Phi(z)| \leq C\,(\text{dist}\,(z, E_1 \cup E_2))^N, \quad |z| < 1. \tag{17}$$

By construction, there is a constant $C_1 > 0$ such that if $z_0 \in E$,

$$|\Phi(z)\,f(z)\,F_N(z)| < C_1\,|z - z_0|^N \text{ for all } |z| = 1. \tag{18}$$

Since the function $C_1(z - z_0)^N$ is outer, this estimate holds for all $|z| < 1$, and the estimate (b) follows.

4. Let I be a closed ideal with boundary zero set E and innerfactor G. To prove Theorem 3, we argue as follows: Let $J = \{f \in I(E, 1): fG \in I\}$. Then Theorem 2 guarantees that J is a closed ideal containing the functions fG^{-1} for all f in I. Thus by the special case of Theorem 3, $J = I(E, 1)$. Now let $g \in I(E, G)$. Then again $gG^{-1} \in I(E, 1)$, so $gG^{-1} \in J$, and thus $g \in I$. Therefore, $I \subseteq I\,(E, G) \subseteq I$ and Theorem 3 is proved.

REFERENCES

1. A. Beurling, On two problems concerning linear transformations in Hilbert space, Acta Math., 81 (1949), 239-255.

2. T. Carleman, L'integral de Fourier et questions qui s'y rattachent, Uppsula, 1935.
3. P. L. Duren, Theory of H^p Spaces. Academic Press, 1970.
4. V. P. Havin, On the factorization of analytic functions smooth on the boundary, Zapiski Nauknikh Seminarov LOMI, 22 (1971).
5. B. I. Korenblyum, Closed ideals in the ring A^n, Func. Anal. and Its Appl., 6, No. 3 (1972), 203-214.
6. A. L. Matheson, Approximation of analytic functions satisfying a Lipschitz condition, to appear.
7. W. Rudin, The closed ideals in an algebra of analytic functions, Canadian J. Math., 9 (1957), 426-434.
8. F. A. Shamoyan, Division by an inner function in some spaces of functions analytic in the disc, Zapiski Nauknikh Seminarov LOMI, 22 (1971), 206-208.
9. B. A. Taylor and D. L. Williams, Ideals in rings of analytic functions with smooth boundary values, Canadian J. Math, 22 (1970), 1266-1283.
10. A. Zygmund, Trigonometric Series. Cambridge University Press, 1959.

A SEPARABLE REFLEXIVE BANACH SPACE HAVING NO FINITE DIMENSIONAL ČEBYŠEV SUBSPACES

Peter Ørno
Ohio State University
Columbus, Ohio 43210

Introduction. A non-dense linear subspace Y of a normed linear space $(X, \|\cdot\|)$ is called Čebyšev if for each $x \in X$ there is a unique $y_0 \in Y$ such that $\|x - y_0\| = \inf\{\|x - y\| : y \in Y\}$. In [6] Singer asked the question: Does every reflexive Banach space have Čebyšev subspaces of every finite dimension? The answer to this question is affirmative for finite dimensional Banach spaces [1]. In this note we renorm $L_2[0,1]$ to have the property stated in the title, thus answering this question in the negative for infinite dimensional spaces. For further terminology and examples in the theory of best approximations, we refer the reader to [5].

In Section 1 we give a sufficient condition for a finite dimensional subspace to be non-Čebyšev, and state an example of a non-separable reflexive space having no separable Čebyšev subspaces.

The main result is given in Section 2. It is known that $L_1[0,1]$ has no finite dimensional Čebyšev subspaces [2], and together with the result of Section 1 this serves as motivation for our construction.

1. Non-Čebyšev subspaces.

Theorem 1. Let $y_1, \cdots, y_n$ be elements of a normed linear space $(X, \|\cdot\|)$. If there exist $x \in X$ and $\epsilon > 0$ such that $\|x + \sum_{k=1}^{n} \delta_k y_k\| = 1$ for all $|\delta_k| \leq \epsilon$ $(1 \leq k \leq n)$, then $Y = \text{span}(y_1, \cdots, y_n)$ is not Čebyšev.

Proof. Note that $\|x\| = 1$ and $x \notin Y$ follow easily from the condition stated in the theorem. Let $y_0 \in Y$, $\|y_0\| = 1$, and let $\lambda_0 \geq 0$ such that $\|x - \lambda_0 y_0\| = \inf\{\|x-y\| : y \in Y\}$. Then $\Phi(\lambda) = \|x - \lambda y_0\|$ defines a convex function for $\lambda \geq 0$ which assumes a minimum at λ_0 and $\Phi(\lambda_0) \leq 1$. But $\|x + \sum_{k=1}^{n} \delta_k y_k\| = 1$ for $|\delta_k| \leq \epsilon$ implies that $\Phi(\lambda) = 1$ in a neighborhood of $\lambda = 0$. Thus by convexity $\Phi(\lambda_0) = 1$. Hence Y is not Čebyšev.

Remark 1. An easy argument shows that Theorem 1 still holds if $|\delta_k| \leq \epsilon$ $(1 \leq k \leq n)$ is replaced by $0 \leq \delta_k \leq \epsilon$ $(1 \leq k \leq n)$.

Remark 2. Using Theorem 1 and Remark 1, it can be shown that $\ell_2(\Gamma)$, Γ uncountable, with the renorming $\|x\| = \max\{\|x\|_\infty, \frac{1}{2}\|x\|_2\}$ has no finite dimensional Čebyšev subspaces and even no separable Čebyšev subspaces. It can also be seen that $(\ell_2(\Gamma), \|\cdot\|)$ has length one segments in all directions on the surface of its unit ball.

2. <u>The Main Result</u>. In the subsequent paragraphs, L_1 and L_2 denote the (real) spaces $L_1[0,1]$ and $L_2[0,1]$, and m denotes Lebesgue measure. For $f \in L_2$ define

$$\|f\| = \max\{\|f\|_1, \tfrac{1}{3}\|f\|_2\},$$

where $\|\cdot\|_1$ and $\|\cdot\|_2$ denote the original norms on L_1 and L_2. Since $\|\cdot\|_1 \leq \|\cdot\|_2$, this norm is equivalent to the original norm on L_2. In order to prove that the separable reflexive space $(L_2, \|\cdot\|)$ has no finite dimensional Čebyšev subspaces, it suffices (by Theorem 1 and Remark 1) to prove the following statement.

<u>Theorem 2</u>. If $g_1, \cdots, g_n \in L_2$ with $\|g_k\|_2 = 1$ $(1 \leq k \leq n)$, then there exist $f \in L_2$ and $\epsilon > 0$ such that $\|f + \sum_{k=1}^{n} \delta_k g_k\| = 1$ for all $0 \leq \delta_k \leq \epsilon$ $(1 \leq k \leq n)$.

<u>Proof</u>. Let $M \geq 0$. For I and J disjoint subsets of $\{1, \cdots, n\}$, define

$$B(I,J) = \{t \mid g_i(t) > M,\ i \in I;\ g_j(t) < -M,\ j \in J;\ |g_k(t)| \leq M,\ k \notin I \cup J\}.$$

Note that $[0,1]$ is the disjoint union $\bigcup\{B(I,J) \mid I \cap J = \emptyset\}$. Using Liapunoff's Theorem ([3] or [4]) let $B_1(I,J)$ and $B_2(I,J)$ be disjoint subsets of $B(I,J)$ satisfying $B_1(I,J) \cup B_2(I,J) = B(I,J)$ and $\int_{B_1(I,J)} g_k\,dm = \int_{B_2(K,J)} g_k\,dm$ $(1 \leq k \leq n)$.

For $c \geq 0$ define

$$f = c(\chi_{B_1(\emptyset,\emptyset)} - \chi_{B_2(\emptyset,\emptyset)})$$
$$+\sum\{(\sum_{k \in I \cup J} |g_k|)(\chi_{B_1(I,J)} - \chi_{B_2(I,J)}) \mid I \cap J = \emptyset,\ I \cup J \neq \emptyset\}.$$

Then it is not difficult to see that there are $M > 0$ and $c > 0$ such that $\|f\|_1 = 1$ and $\|f\|_2 < 2$.

Choose $0 < \epsilon \leq \min\{\frac{c}{n\cdot M}, \frac{2 - \|f\|_2}{n}\}$. Then $\epsilon \leq \frac{1}{n}$, and for $0 \leq \delta_k \leq \epsilon$ $(1 \leq k \leq n)$ we have $\|f + \sum_{k=1}^{n} \delta_k g_k\|_2 \leq 2$. So $\|f\| = 1$ and $\frac{1}{3}\|f + \sum_{k=1}^{n} \delta_k g_k\|_2 < 1$ for $0 \leq \delta_k \leq \epsilon$. Now $\|f + \sum_{k=1}^{n} \delta_k g_k\|_1 = \sum_{I \cap J = \emptyset} \int_{B(I,J)} |f + \sum_{k=1}^{n} \delta_k g_k|\,dm$, so computing term-by-term we have

$$\int_{B(\emptyset,\emptyset)} |f + \sum_{k=1}^{n} \delta_k g_k|\,dm = \int_{B_1(\emptyset,\emptyset)} (c + \sum_{k=1}^{n} \delta_k g_k)\,dm + \int_{B_2(\emptyset,\emptyset)} (c - \sum_{k=1}^{n} \delta_k g_k)\,dm,$$

since $|\sum_{k=1}^{n} \delta_k g_k| \leq n\cdot\epsilon\cdot M \leq c$ on $B(\emptyset,\emptyset)$. And for $I \cup J \neq \emptyset$,

$$\int_{B(I,J)} |f + \sum_{k=1}^{n} \delta_k g_k| dm = \int_{B_1(I,J)} \left(\sum_{k \in I} (1 + \delta_k) g_k - \sum_{k \in J} (1 - \delta_k) g_k + \sum_{k \notin I \cup J} \delta_k g_k \right) dm$$

$$+ \int_{B_2(I,J)} \left(\sum_{k \in I} (1 - \delta_k) g_k - \sum_{k \in J} (1 + \delta_k) g_k - \sum_{k \notin I \cup J} \delta_k g_k \right) dm,$$

since $\sum_{k \in I} (1 + \delta_k) g_k + \sum_{k \in J} (-1 + \delta_k) g_k + \sum_{k \notin I \cup J} \delta_k g_k \geq (1 - n\epsilon) M \geq 0$ on $B_1(I,J)$,

and $\sum_{k \in I} (-1 + \delta_k) g_k + \sum_{k \in J} (1 + \delta_k) g_k + \sum_{k \notin I \cup J} \delta_k g_k \leq (-1 + n\epsilon) M \leq 0$ on $B_2(I,J)$.

The $\delta_k g_k$ terms in each pair of $B_1(K,J)$, $B_2(I,J)$ integrals cancel, thus $\|f + \sum_{k=1}^{n} \delta_k g_k\|_1 = \|f\|_1 = 1$. Hence $\|f + \sum_{k=1}^{n} \delta_k g_k\| = 1$ for $0 \leq \delta_k \leq \epsilon$ $(1 \leq k \leq n)$. This completes the proof.

Remark 3. By modifying the above proof (take $n = 1$, $\|g\|_1 = 1$ and $M = 0$) we note that L_1 has length one segments in all directions on its sphere. A similar property holds for the space $(L_2, \|\cdot\|)$ constructed above.

Theorem 3. $(L_2, \|\cdot\|)$ has segments of length at least $\frac{1}{12}$ in all directions on its sphere.

Proof. Let $g \in L_2$, $\|g\|_2 = 1$. Construct the sets $B(I,J)$ as in the proof of Theorem 2 for the case $n = 1$, $M = 2$ and $g_1 = g$. Then $m(B(\emptyset,\emptyset)) \geq \frac{3}{4}$. However, define f by

$$f = c(\chi_{B_1(\emptyset,\emptyset)} - \chi_{B_2(\emptyset,\emptyset)}) + \frac{1}{2} g (\chi_{B_1(\{1\},\emptyset) \cup B_2(\emptyset,\{1\})} - \chi_{B_1(\emptyset,\{1\}) \cup B_2(\{1\},\emptyset)}),$$

where c is chosen so that $\|f\|_1 = 1$, thus $\frac{1}{2} \leq c \leq \frac{4}{3}$. Then $\|f\|_2 < 2$. The computations in the proof of Theorem 2 show that $\|f + \delta g\| = 1$ if $0 \leq \delta \leq \frac{1}{4} \leq \frac{c}{M}$. And $\frac{1}{4}\|g\| \geq \frac{1}{12}\|g\|_2 = \frac{1}{12}$.

REFERENCES

1. G. Ewald, D. G. Larman, C. A. Rogers, The directions of the line segments and of the n-dimensional balls on the boundary of a convex body in Euclidean space. Mathematika, 17(1970), 1-20.
2. M. G. Krein, L-problems in abstract linear spaces. D.N.T.V.U. Karkov (1938), 171-199.
3. A. A. Liapunoff, On completely additive vector-functions. Izv. Akad. Nauk S.S.S.R., 4(1940), 465-478.
4. J. L. Lindenstrauss, A short proof of Liapunoff's convexity theorem. J. Math. Mech., 15(1966), 971-972.
5. I. Singer, Best Approximations in Normed Linear Spaces by Elements of Linear Subspaces. Die Grundlehren der mathematischen Wissenschaften, Band 171, Springer-Verlag, Berlin-Heidelberg, 1970.
6. I. Singer, On best approximations in normed linear spaces by elements of subspaces of finite codimension. Rev. Roum. math. pures et appl., 17(1972), 1245-1256.

A NONLOCALLY CONVEX F-SPACE WITH THE HAHN-BANACH APPROXIMATION PROPERTY

James W. Roberts
Department of Mathematics
University of South Carolina
Columbia, SC 29208

1. Introduction and Terminology. An F-space X is a topological vector space whose topology is given by a complete invariant metric. X^* is the set of continuous linear functionals on X and the weak topology on X is the weakest topology so that all the linear functionals on X are continuous. X is said to have the Hahn-Banach approximation property if the weak closure of every proper closed subspace is proper. If X is locally convex, then X clearly has the Hahn-Banach approximation property. A natural question is the following: if X has the Hahn-Banach approximation property, then is X locally convex? N. J. Kalton has shown that if X is a separable F-space with the Hahn-Banach approximation property and X^* separates points in X, then X is locally convex [1]. In this paper we shall construct a nonlocally convex F-space X which has the Hahn-Banach approximation property. The space X will be separable but X^* will fail to separate points in X. For a discussion of this problem and related questions and results concerning F-spaces see J. H. Shapiro [5].

If X is a linear invariant semimetric space with invariant semi-metric d, then for every $x \in X$, $N(x) = d(x,0)$ is called a paranorm. If $N(x) = 0$ implies $x = 0$ (d is a metric), then N is said to be total. If $0 < p < 1$, N is called a p-norm if for every $x \in X$ and $\alpha \in R$. $N(\alpha x) = |\alpha|^p N(x)$. An example is the space of measurable function f on [0,1] such that

$$N_p(f) = \int_0^1 |f(x)|^p dx < \infty.$$

N_p is the p-norm on $L^p([0,1])$. If E is a set in a linear invariant metric space S, then [E] will denote the closed convex hull of E. Also $[E_1, \cdots, E_n]$ will denote $[\bigcup_{i=1}^n E_i]$. A point x_0 in X is called a needle point if $x_0 \neq 0$ and for every $\delta > 0$, there exists a finite set F in X such that

(1) $x_0 \in [F]$.

(2) if $x \in F$, then $N(x) < \delta$.

(3) if $x \in [0,F] = [\{0\}, F]$, then there exists $\alpha \in [0,1]$ such that $N(x-\alpha x_0) < \delta$.

The set F is called a δ-needle set for x_0. The space X is called a needle point space if every nonzero point in X is a needle point. The spaces $L^p([0,1])$, $0 < p < 1$, are needle point spaces (see Roberts [3], [4]).

Finally, the author was informed by J. H. Shapiro that N. J. Kalton has obtained a similar example.

2. The Construction of the F-space. To construct the space we shall first construct some fairly pathological paranorms on finite dimensional spaces. The finite dimensional spaces will then be pieced together so that they are almost in direct sum, i.e. their quotient by a one dimensional space will be in direct sum. First we shall need a technical lemma which appears in [2] as Proposition 2.1.

Lemma 2.1: Let V be a finite dimensional space, K a compact subset of V which spans V, and ϕ a function from the nonnegative reals to the nonnegative reals such that

(1) $0 = \phi(0) = \lim_{\alpha \to 0} \phi(\alpha)$,

(2) ϕ is monotone increasing, and

(3) there exist $\varepsilon > 0$ and $c > 0$ such that if $0 \leq \alpha < \varepsilon$, then $\phi(\alpha) \geq c\alpha$.

If for every $x \in V$ we let $N(x) = \inf\{\sum_{i=1}^{n} \phi(|\alpha_i|): \alpha_i \in R,\ x_i \in K,\ \sum_{i=1}^{n} \alpha_i x_i = x\}$, then N is a total paranorm on V.

Theorem 2.2: If $0 < p < 1$ and $\delta > 0$, then there exists a finite dimensional space V with a p-norm N, a point $x_0 \in V$ and a finite set $F \subset V$ such that

(1) $N(x_0) = 1$.

(2) F is a δ-needle set for x_0.

(3) There exists a pseudonorm $\|\cdot\|$ on V such that if N' is the Rx_0-quotient paranorm on V induced by N, then $N'(x) = \|x\|^p$ for every $x \in V$.

Proof: Let x_0 be a point in $L^p([0,1])$ such that $N_p(x_0) = 1$, where N_p denotes the p-norm on $L^p([0,1])$. Since $L^p([0,1])$ is a needle point space there exists a $\delta/2$-needle set F about x_0. Let V = span F and let

$$E = \{y - \alpha x_0 : y \in [-F, F],\ \alpha \in [-1,1], \text{ and } N_p(y - \alpha x_0) \leq \delta\}.$$

Define a new paranorm N_1 on V by

$$N_1(x) = \delta \inf\{\sum_{i=1}^{n} |\alpha_i|^p : \sum_{i=1}^{n} \alpha_i y_i = x,\ y_i \in E,\ \alpha_i \in R\}.$$

By Lemma 2.1 N_1 is a total paranorm on V. Since $-E = E$, N_1 can be expressed by the following:

$$N_1(x) = \delta \inf\{\sum_{i=1}^{n} \alpha_i^p : \sum_{i=1}^{n} \alpha_i y_i,\ y_i \in E,\ \alpha_1 \geq 0\}.$$

Observe further than N_1 is a p-norm. Now suppose $y_1, \cdots, y_n \in E$, $\alpha_1, \cdots, \alpha_n$ are nonnegative and $\sum_{i=1}^{n} \alpha_i y_i = x_0$. Since each $y_i \in E$, $N_p(y_i) \leq \delta$. Therefore

$$1 = N_p(x_0) \leq \sum_{i=1}^{n} \alpha_i^p N_p(y_i) \leq \delta \sum_{i=1}^{n} \alpha_i^p.$$

It follows that $1 \leq N_1(x_0)$. Define $N(x) = N_1(x_0)^{-1}N_1(x)$ for every $x \in X$. Thus N is a p-norm on V, $N(x_0) = 1$, $N(x) \leq N_1(x)$ for every $x \in V$, and if $x \in E$, then $N(x) < \delta$. If $x \in F$, then $N_p(x) < \delta$ and therefore $x \in E$. Thus $N(x) < \delta$. If $x \in [0,F]$, then there exists $\alpha \in [0,1]$ such that $N_p(x-\alpha x_0) > \delta$. Hence $x-\alpha x_0 \in E$ and we have $N(x-\alpha x_0) \leq \delta$. Thus F is a δ-needle set about x_0 with respect to the paranorm N. Thus conditions (1) and (2) are satisfied.

We now show that condition (3) holds. Let $\lambda = \delta N_1(x_0)^{-1}$ and let

(2.1) $$\|x\| = \lambda^{1/p} \inf \{\alpha: x \in \alpha U, \alpha > 0\} \text{ where } U = [-F, F] + Rx_0.$$

Since U is convex, symmetric and absorbing, $\|\cdot\|$ is a pseudonorm. Observe that if $y \in [-F, F]$, then $y = y_1 - y_2$ where $y_1, y_2 \in [0,F]$. There exist $\beta_1, \beta_2 \in [0,1]$ such that $N_p(y_1-\beta_1 x_0) < \delta/2$ and $N_p(y_2-\beta_2 x_2) < \delta/2$. Thus if $\beta = \beta_1-\beta_2 \in [-1,1]$ then, $N_p(y-\beta x_0) < \delta$, i.e. if $y \in [-F,F]$, then there exists $\beta \in [-1,1]$ such that $y - \beta x_0 \in E$. Now suppose $x \in V$ and $\epsilon > 0$. There exists $\alpha > 0$ such that $x \in \alpha U$ and $\epsilon + \|x\|^p > \lambda\alpha^p$. Then $x = \alpha(\alpha_0 x_0 + y)$ where $\alpha_0 \in R$ and $y \in [-F,F]$. There exists $\beta \in [-1,1]$ such that $y - \beta x_0 \in E$. Now $x = \alpha(y-\beta_0 x_0) + \alpha(\alpha_0 + \beta_0)x_0$. Hence,

$$N'(x) = N'(\alpha(y-\beta_0 x_0)) \leq N(\alpha(y-\beta_0 x_0)) = \lambda\alpha^p < \|x\|^p + \epsilon.$$

Thus $N'(x) \leq \|x\|^p$.

Once again let $x \in V$ and let $\epsilon > 0$. There exists $\alpha \in R$ (with $x \neq \alpha x_0$) such that $\epsilon/2 + N'(x) \geq N(x-\alpha x_0)$ and there exist $y_1, \cdots, y_n \in [-F, F]$, $\beta_1, \cdots, \beta_n \in [-1, 1]$, and $\alpha_1, \cdots, \alpha_n$ nonnegative such that $y_i-\beta_i x_0 \in E$, $\epsilon/2 + N(x-\alpha x_0) \geq \lambda \sum_{i=1}^{n} \alpha_i^p$ and $\sum_{i=1}^{n} \alpha_i (y_i-\beta_i x_0) = x - \alpha x_0$. Thus

$$x = \sum_{i=1}^{n} \alpha_i (y_i - (\beta_i - \alpha/\sum_{i=1}^{n} \alpha_i)x_0).$$

It is clear from (2.1) that for each i, $1 \leq i \leq n$,

$$\|y_i\| = \|y_i - (\beta_i - \alpha/\sum_{i=1}^{n} \alpha_i)x_0\| = \|y_i - \beta_i x_0\|$$

and

$$\|y_i\|^p \leq \lambda.$$

Hence

$$\|x\|^p \leq (\sum_{i=1}^{n} \alpha_i\|y_i\|)^p \leq \sum_{i=1}^{n} \alpha_i^p\lambda \leq N(x-\alpha x_0) + \epsilon/2 \leq N'(x) + \epsilon.$$

It follows that $\|x\|^p \leq N'(x)$ and therefore $\|x\|^p = N'(x)$ for every $x \in V$.

To construct the space we shall use an infimum of a family of paranorms. If V is a vector space spanned by a sequence V_i of subspaces and each V_i has a paranorm N_i we define $N = \inf \{N_i: i = 1, 2, \cdots\}$by

$$N(x) = \inf \{\sum_{i=1}^{n} N_i(x_i): x_i \in V_i, \sum_{i=1}^{n} x_i = x\}.$$

N is a paranorm (see Proposition 2.3 in [2]) and N is the largest paranorm smaller than each N_i on V_i.

Theorem 2.3: There exists an F-space X such that X contains a needle point x_0 and X/Rx_0 is a Banach space.

Proof: Choose a sequence $p(n)$, $0 < p(n) < 1$, such that $\alpha^{p(n)} \leq 2\alpha$ if $\alpha \geq 2^{-n}$. Note that $\lim p(n) = 1$. Also let $\delta(n)$ be a sequence of positive numbers such that $\lim_{n\to\infty} \delta(n) = 0$. Now let W be a countably infinite dimensional vector space over the real numbers and let x_0 be a nonzero vector in W. One can easily construct a sequence V_n of finite dimensional subspaces in W with $p(n)$ - norms N_n such that

(1) If $1 \leq i \leq n$, then $V_i \cap \Sigma \{V_j: 1 \leq j \leq n, j \neq i\}) = Rx_0$.

(2) V_n contains a finite set F_n such that F_n is a $\delta(n)$ - needle set for x_0 and $N_n(x_0) = 1$.

(3) For every n, there exists a pseudonorm $\|\cdot\|_n$ on V_n such that if N_n' is the $V_n' = V_n/Rx$ - quotient paranorm, then $N_n'(x) = \|x\|_n^{p(n)}$ for every $x \in V_n$.

An equivalent formulation of condition (1) is that the vector spaces $V_1', \cdots, V_n'$ are in direct sum. The construction is carried out inductively on n as follows: if $V_1, \cdots, V_n$ and $N_1, \cdots, N_n$ have been constructed, choose V_{n+1} in W of suitable dimension so that $V_{n+1} \cap (\sum_{i=1}^{n} V_i) = Rx_0$ and V_{n+1} has a $p(n+1)$ - norm N_{n+1} satisfying conditions (2) and (3) obtained by applying Theorem 2.2. The construction of V_1 and N_1 is obvious. Let V be the span of all the spaces $V_1, V_2, \cdots$ and define the paranorm $N = \inf \{N_n: n = 1, 2, \cdots\}$, i.e.,

$$N(x) = \inf \{\sum_{i=1}^{n} N_i(x_i): \sum_{i=1}^{n} x_i = x, x_i \in V_i\}$$

for all $x \in V$. Let $V' = V/Rx_0$ and let N' denote the V/Rx_0 - quotient paranorm.

If $x_i \in V_i$, $1 \leq i \leq n$, and $\sum_{i=1}^{n} x_i = x_0$, then each $x_i \in Rx_0$, i.e. $x_i = \alpha_i x_0$ for some $\alpha_i \in R$ and $\sum_{i=1}^{n} \alpha_i = 1$. Thus $\sum_{i=1}^{n} N_i(x_i) = \sum_{i=1}^{n} |\alpha_i|^{p(i)} \geq 1$. Hence $N(x_0) = 1$. If $x \in V_n$, then $N(x) \leq N_n(x)$. Since F_n is a $\delta(n)$ - needle set in V_n with respect to N_n it is also a $\delta(n)$ - needle set with respect to N. Hence x_0 is a needle point in V.

We now claim that if $x_i \in V_i$, $1 \leq i \leq n$, then

(2.2)
$$N'(\sum_{i=1}^{n} x_i) = \sum_{i=1}^{n} N_i'(x_i).$$

Let $x = \sum_{i=1}^{n} x_i$, $x_i \in V_i$, $1 \leq i \leq n$ and suppose $\epsilon > 0$. Then there exists $\alpha \in R$ such that $N'(x) + \epsilon > N(x-\alpha x_0)$. There exists $y_i \in V_i$, $1 \leq i \leq m$ (with $m \geq n$) such that $\sum_{i=1}^{m} y_i = x-\alpha x_0$ and $N'(x) + \epsilon > \sum_{i=1}^{m} N_i(y_i)$. For $n < i \leq m$, $y_i \in Rx_0$. If $1 \leq i \leq n$, then $y_i = x_i + \alpha_i x_0$ for $\alpha_i \in R$. Thus $N'(x) + \epsilon > \sum_{i=1}^{n} N_i(x_i+\alpha_i x_0) + \sum_{i=n+1}^{m} N_i(y_i) \geq \sum_{i=1}^{n} N_i'(x_i+\alpha_i x_0) + \sum_{i=n+1}^{m} N_i'(y_i) = \sum_{i=1}^{n} N_i'(x_i)$. Since $\epsilon > 0$ is arbitrary, $N'(\sum_{i=1}^{n} x_i) \geq \sum_{i=1}^{n} N_i'(x_i)$. As a special case, if $x \in V_n$, then $N'(x) \geq N_n'(x)$. Since $N \leq N_n$ on V_n, $N'(x) \leq N_n'(x)$. Thus $N'(x) = N_n'(x)$. Now if $x_i \in V_i$, $1 \leq i \leq n$, then

$$N'(\sum_{i=1}^{n} x_i) \leq \sum_{i=1}^{n} N'(x_i) = \sum_{i=1}^{n} N_i'(x_i).$$

Thus $N'(\sum_{i=1}^{n} x_i) = \sum_{i=1}^{n} N'(x_i)$. Now let X denote the completion of V with respect to N and X' the completion of V' with respect to N' (N and N' will still denote the paranorms on X and X', respectively). Of course, $X/Rx_0 = X'$ and N' is the Rx_0 - quotient paranorm on X. By (2.2) it is easy to see that X' is the ℓ^1-sum of the spaces (V_n', N_n'), i.e. $x \in X'$ if and only if there exist $x_i \in V_i'$ such that $\sum_{i=1}^{\infty} N_i'(x) < \infty$ and $x = \sum_{i=1}^{\infty} x_i$. Also $N'(x) = \sum_{i=1}^{\infty} N_i'(x_i)$.

We now show that X' is a Banach space. Suppose $x \in X'$ and $N'(x) \leq 1$. Then $x = \sum_{i=1}^{\infty} x_i$ with $x_i \in V_i'$ and

$$N'(x) = \sum_{i=1}^{\infty} N_i'(x_i) = \sum_{i=1}^{\infty} \|x_i\|_i^{p(i)}.$$

Since $\|x_i\|_i^{p(i)} = N_i'(x_i) \leq 1$, $\|x_i\|_i^{p(i)} \geq \|x_i\|_i$. Hence $\sum_{i=1}^{\infty} \|x_i\|_i \leq N'(x)$. Thus if $x \in X'$ and $x = \sum_{i=1}^{\infty} x_i$ where $x_i \in V_i'$ we may define $\|x\| = \sum_{n=1}^{\infty} \|x_n\|_n$ since clearly $\sum_{n=1}^{\infty} \|x_n\|_n < \infty$. Furthermore the identity map from (X', N') to $(X', \|\cdot\|)$ is continuous. To complete the proof we must show that $(X', \|\cdot\|)$, is a Banach space, i.e. X' is complete. Suppose $x_n \in V_n'$ and $\sum_{n=1}^{\infty} \|x_n\|_n < \infty$. By our choice of p(n), $\|x_n\|_n^{p(n)n} \leq 2\|x_n\|_n + 2^{-p(n)n}$. Since $\lim_{n\to\infty} p(n) = 1$, $\sum_{n=1}^{\infty} 2^{-p(n)n} < \infty$. Hence $\sum_{n=1}^{\infty} N_n'(x_n) < \infty$. Thus X' is the ℓ^1-sum of the spaces $(V_n', \|\cdot\|_n)$ and $(X', \|\cdot\|)$ is a Banach space. By the Open Mapping Theorem N' and $\|\cdot\|$ are equivalent on X' and (X', N') is a Banach space.

Observe now that if x_0 is a needle point in a space X, T is a continuous linear operator from X to another F-space Y and $T(x_0) \neq 0$, then $T(x_0)$ is also a needle point. Thus if Y is locally convex then $T(x_0) = 0$ (since it contains no needle points). As a consequence, if $\ell \in X^*$ and x_0 is a needle point of X, then $\ell(x_0) = 0$. We now show that the space X constructed in Theorem 2.3 has the Hahn-Banach approximation property.

Theorem 2.4: There exists an F-space X such that X is not locally convex and X has the Hahn-Banach approximation property.

Proof: Let X be an F-space containing a needle point x_0 such that X/Rx_0 is a Banach space. We claim that a closed subspace Y of X is weakly closed if and only if $x_0 \in Y$. Suppose $x_0 \in Y$. Then the map

$$\phi: \ X/Rx_0 \to X/Y$$

is a continuous open map. Since X/Rx_0 is a Banach space, X/Y is locally convex (actually another Banach space). But then Y is weakly closed. If ℓ is a continuous linear functional on X, then $\ell(x_0) = 0$. It follows that if Y is weakly closed, then $x_0 \in Y$.

Now uppose that Y is a proper closed subspace of X. Then $Y + Rx_0$ is weakly closed. If $Y + Rx_0 = X$, then Y is a closed subspace of codimension one and therefore Y is weakly closed, i.e. $Rx_0 \subset Y$ and $Y = Y + Rx_0 = X$. This is a contradiction since Y is proper. Thus $Y + Rx_0$ is weakly closed and proper.

Note: $Y + Rx_0$ is, of course, the weak closure of Y.

REFERENCES

1. N. J. Kalton, Quotients of F-spaces, to appear.
2. J. W. Roberts, A compact convex set with no extreme points, Studia Math., to appear.
3. ________, Pathological compact convex sets in the spaces $L^p([0,1])$, $0 \leq p < 1$, Altgeld Book, to appear.
4. ________, Compact convex sets with no extreme points in the spaces $L^p([0,1])$, $0 \leq p < 1$, to appear.
5. J. H. Shapiro, Remarks on F-spaces of analytic functions to appear.

THE BANACH-MAZUR DISTANCE BETWEEN FUNCTION ALGEBRAS ON DEGENERATING RIEMANN SURFACES

Richard Rochberg[1]
Department of Mathematics
Washington University
St. Louis, MO 63130

1. Introduction and Summary. The Banach-Mazur distance between algebras of continuous boundary value analytic functions on finite bordered Riemann surfaces induces a metric on the Riemann space of such surfaces and this metric induces the classical moduli topology. In this paper we show that the metric space so obtained is not complete. We do this by exhibiting three types of geometric degeneration of Riemann surfaces which yield Cauchy sequences with no limit points. Two of the types of degeneration are the analogs of types of degeneration of compact surfaces known as "pinching a waist" and "pinching a handle". The third type of degeneration, "pinching a boundary", has no analog for compact surfaces. In each case the construction actually exhibits an algebra of analytic functions to which the Cauchy sequence converges. For the first two types of degeneration, the limit algebras are algebras of all continuous boundary value analytic functions on degenerate Riemann surfaces; in the third case, the limit algebra is a subalgebra of finite codimension of an algebra of all continuous boundary value analytic functions on a Riemann surface.

It is a tempting conjecture that repeated application to these three types of degenerative processes yield the full Cauchy completion of the Riemann space. Although this conjuncture appears difficult, there is some evidence for it [8], [9].

Two other consequences of these constructions are worth noting. First, for the function algebras considered the homeomorphism type of the maximal ideal space is not stable under infinitesimal perturbations of the algebra in the Banach-Mazur distance. This is an interesting contrast to the fact that, under suitable restrictions, the homeomorphism type of the Shilov boundary is stable under such perturbations. Second, the constructions provide many of the pieces for a proof of the folklore theorem that any two algebras of all continuous boundary value analytic functions on finite bordered Riemann surfaces are isomorphic as Banach spaces.

Section 2 contains definitions and notation. The tyree types of degeneration are presented in Sections 3, 4, and 5. Consequences of these constructions are in Section 6. The final section contains some open questions.

The general study of Banach-Mazur limits of function algebras uses a number of the deeper results from the general theory of function algebras. In contrast, the

[1]This work supported in part by NSF Grant MPS - 7506367.

constructions we will present are based on classical function theory and elementary functional analysis.

A general reference for Riemann surfaces is [11].

2. <u>Notation and Terminology</u>. Let $\mathcal{S}$ be the class of finite bordered Riemann surfaces. For integers g and k, $g \geq 0$, $k \geq 1$, let $\mathcal{S}(g,k)$ be the class of elements in $\mathcal{S}$ of genus g which have k boundary components. For S in $\mathcal{S}$ we denote the border of S by ∂S and the interior of S, $S\setminus\partial S$, by int(S). For the same g and k let $R(g,k)$ be the Riemann space of $\mathcal{S}(g,k)$, the space of conformal equivalence classes of elements of $\mathcal{S}(g,k)$. We will regard the assumption that g and k are integers, $g \geq 0$, $k \geq 1$, as an implicit part of the notation $\mathcal{S}(g,k)$ and $R(g,k)$ and we will often denote elements of $R(g,k)$ by representatives in $\mathcal{S}(g,k)$. For S in $\mathcal{S}$, denote by A(S) the supremum normed Banach algebra of all functions continuous on S and analytic on int(S).

For Banach spaces A and B let L(A,B) be the set of continuous invertible linear maps from A onto B. The Banach-Mazur distance between A and B is defined by

$$d(A,B) = \log \inf\{\|T\| \, \|T^{-1}\| \ ; \ T \in L(A,B)\}.$$

Define a function d on $R(g,k)$ by $d(S,S') = d(A(S),A(S'))$ where $S, S' \in \mathcal{S}(g,k)$. It is known [7] that d is a metric on $R(g,k)$ and that the topology induced by this metric is equivalent to that induced by the Teichmüller metric.

The examples we will be constructing will be elements of the Cauchy completion of the metric spaces $(R(g,k),d)$. The examples can be understood as starting with an S in some $\mathcal{S}(g,k)$ and a closed non-contractible curve γ on S and "pinching γ to a point". A precise formulation of this idea can be given using the notion of harmonic length of a curve introduced by Landau and Osserman [3]. Let γ be a smooth closed curve on S. Define the harmonic length of γ, $\ell(\gamma)$ by

$$\ell(\gamma) = \sup\{\int_\gamma {}^*du; \text{ u a real harmonic function on S with } \sup|u| \leq 1\}.$$

(Here $^*du$ denotes the harmonic conjugate of the differential du.)

3. <u>Pinching a Waist</u>. This example can be interpreted geometrically as starting with a surface S in $\mathcal{S}$ and a simple closed curve γ on S which separates S and is not homologous to zero and letting the harmonic length of γ go to zero. This procedure is often referred to as "pinching a waist". A discussion of this type of degeneration for compact surfaces is given by Lebowitz in [4].

For i = 1,2 select the following objects: integers g_i, k_i with $g_i \geq 0$, $k_i \geq 1$; S_i in $\mathcal{S}(g_i,k_i)$, p_i points of $S_i\setminus\partial S_i$, f_i in $A_i = A(S_i)$ $\|f_i\| = 1$, f_i has a simple zero at p_i and no other zeros on S_i. All these objects are fixed for the remainder of the section. There is an $\epsilon_0 > 0$ such that for $0 < \epsilon < \epsilon_0$ the sets $\{|f_i| = \epsilon\}$ are simple closed curves on which the change of $\arg(f_i)$ is 2π. For all ϵ with $0<\epsilon<\epsilon_0$

we will describe the construction of a surface S_ϵ in $S(g_1+g_2,k_1+k_2)$. For ϵ, $0<\epsilon<\epsilon_0$, for $i=1,2$ let $S_{i\epsilon}=\{q\in S_i; |f_i(q)|\geq\epsilon\}$ and let $\gamma_{i\epsilon}=\{q\in S_i; |f_i(q)|=\epsilon\}$. Let S_ϵ be the Riemann surface obtained by joining $S_{1\epsilon}$ and $S_{2\epsilon}$ along $\gamma_{i\epsilon}$. That is, let γ_ϵ be the set of points obtained by identifying the point of $\gamma_{1\epsilon}$ at which $f_{1\epsilon}$ takes the value $\epsilon e^{i\theta}$ $0\leq\theta<2\pi$, with the point on $\gamma_{2\epsilon}$ at which $f_{2\epsilon}$ takes the value $\epsilon^{e-i\theta}$. As a point set, $S_\epsilon=(S_{1\epsilon}\setminus\gamma_{1\epsilon})\cup\gamma_\epsilon\cup(S_{2\epsilon}\setminus\gamma_{2\epsilon})$. S_ϵ is given the obvious analytic structure and is an element of $S(g_1+g_2,k_1+k_1)$. The verification of this fact is the same as in [4]. Let S_0 be the degenerate Riemann surface obtained from $S_1\cup S_2$ by identifying the points p_1 and p_2 to a single point p^*. Let A_0 be the algebra of functions continuous on S_0 and analytic on the interior of S_0. Geometrically, $\ell(\gamma_\epsilon)$, the length of γ_ϵ, tends to 0 with ϵ and in the limit γ_ϵ collapses to the point p^*. For $0<\epsilon<\epsilon_0$ let $A_\epsilon=A(S_\epsilon)$.

<u>Proposition 1</u>: $d(A_\epsilon,A_0)=O(\epsilon)$.

<u>Proof</u>: It suffices to construct $T_\epsilon\in L(A_0,A_\epsilon)$ with $\|T_\epsilon\|\,\|T_\epsilon^{-1}\|=1+O(\epsilon)$. Decompose A_0 as $A_0=C\oplus A_{1,0}\oplus A_{2,0}$ where $A_{1,0}$ is the set of functions in A_0 which vanish on S_2 and $A_{2,0}$ the functions which vanish on S_1. Thus, for g in A_0 $g=g(p^*)+g_1+g_2$ where g_1 vanishes on S_2 and g_2 vanishes on S_1. This decomposition is clearly a Banach space direct sum. We will denote f_1 by z and f_2 by w. By extending z to be zero on S_2 we can regard z as an element of $A_{1,0}$. Similarly w is regarded as an element of $A_{2,0}$. For $a\in C$ set $T_\epsilon(a)=a$. For g in $A_{1,0}$ define Tg by $(Tg)(p)=g(p)$ for p in S_1, $(Tg)(p)=g(z^{-1}(\frac{\epsilon^2}{w(p)}))$ for p in S_2. Similarly, for h in $A_{2,0}$ let $(Th)(p)=h(p)$ for p in S_2, $(Th)(p)=h(w^{-1}(\frac{\epsilon^2}{z(p)}))$ for p in S_1. Using the direct sum decomposition of A_0, T is extended by linearity to be defined on all of A_0. For f in A_0, Tf is continuous on S_ϵ and analytic except possibly on γ_ϵ. Thus Tf is analytic and hence in A_ϵ. T_ϵ is a linear map of A_0 into A_ϵ and $T_\epsilon(1)=1$.

We now estimate the norm of T_ϵ. Let g be an element of A_0, $\|g\|=1$, $g=c+g_1+g_2$ with g_i in $A_{i,0}$ $i=1,2$. Let p be a point of ∂S_ϵ which came from a point of ∂S_1. Thus

$$(3.1)\qquad |T_\epsilon g(p)|=|T_\epsilon(c+g_1+g_2)(p)|=|c+g_1(p)+g_2(w^{-1}(\epsilon^2/z(p)))|.$$

Since $\|g\|=1$, $\|g_1+g_2\|\leq 2$ and hence $\|g_2\|\leq 2$. By Schwarz's lemma applied to the function g_2 on the disk $|w|<\epsilon_0$, $|g_2(w^{-1}(\epsilon^2/z(p)))|\leq 2\epsilon/\epsilon_0$. Using this with (3.1), we find

$$(3.2)\qquad \big||T_\epsilon g(p)|-|c+g_1(p)|\big|\leq 2\epsilon/\epsilon_0.$$

Since $g\in A_0$, $\|g\|=1$ either there is a point p in ∂S_1, for which $|c+g_1(p)|=1$ or a point p' in ∂S_2 for which $|c+g_2(p')|=1$. In the former case the previous inequality yields

(3.3) $$\|T_\varepsilon g\| \geq 1 - 2\varepsilon/\varepsilon_0.$$

In the latter case, similar estimates yield the same inequality. Since $\|g\| = \max (\sup_{\partial S_1} |c + g_1|, \sup_{\partial S_2} |c + g_2|)$, (3.2) also implies

(3.4) $$\|T_\varepsilon g\| \leq 1 + 2\varepsilon/\varepsilon_0.$$

By (3.3) and (3.4) T_ε is a continuous one to one map of A_0 onto a closed subspace of A_ε. We now show that subspace is all of A_ε. Pick F in A_ε. We denote by z the function $T_\varepsilon(z) \in A_\varepsilon$. Thus z is a coordinate parameter on the annulus $R = \{p \in S_\varepsilon;\ \frac{1}{2}\varepsilon < |z(p)| < \varepsilon_0\}$. On this annulus F can be represented by a Laurant series in z, $F = \sum_1^\infty A_n z^n + C + \sum_1^\infty B_n z^{-n}$. The series $C + \sum B_n z^{-n}$ converges uniformly in $|z| > \frac{2}{3}\varepsilon$ to a function F_1. Hence $F - F_1$ is a function on $S_{1\varepsilon}$ which extends to a function G in $A(S_1)$ which vanishes at p_1. We can regard G as an element of $A_{1,0}$. $T_\varepsilon(G)$ is an element of A_ε which agrees with $F - F_1$ on R. Thus $F - F_1$ is the Laurant series of an element of $T_\varepsilon(A_0)$. Similarly, F_1 is in $T_\varepsilon(A_0)$. Thus F is in $T_\varepsilon(A_0)$. F was arbitrary, hence T_ε is onto. Thus T_ε is in $L(A_0, A_\varepsilon)$. Hence $\|T_\varepsilon\|\ \|T_\varepsilon^{-1}\|$ can be estimated using (3.3) and (3.4). The proposition follows.

4. Pinching a Handle. This example can be interpreted geometrically as starting with a surface S in S of positive genus and a simple closed curve γ on S which does not separate S and which is not homologically trivial. As the harmonic length of γ goes to zero, the handle which supports γ is deformed. In the limit, this handle becomes a point and the genus of the associated Riemann surface is lowered by one. The resulting object is a degenerate Riemann surface which can be realized as an element of S on which two points have been identified.

We will do the construction for the case g = 1, k = 1. The extension to general k would require more notation but no new ideas. The extension to higher g would require more machinery and has not been investigated.

Let D be the unit disk $\{|z| \leq 1\}$. For ε with $0 < \varepsilon < \frac{1}{10}$, let D_ε be D with the two small disks $\{|z| < \varepsilon\}$ and $\{|z - \frac{1}{2}| < \varepsilon\}$ removed. Let S_ε be the surface obtained by identifying the two small boundary circles of D_ε as follows: identify the point z on $\{|z| = \varepsilon\}$ with the point $\frac{1}{2} - \frac{\varepsilon^2}{z}$ on $\{|z - \frac{1}{2}| = \varepsilon\}$. This surface is given the obvious analytic structure and is an element of S(1,1), see [4] for details. The curve γ_ε on S_ε which is formed by identifying the two small circles has the property that $\ell(\gamma_\varepsilon)$ tends to 0 with ε. Let $A_\varepsilon = A(S_\varepsilon)$, $\hat{A}_\varepsilon = A(D_\varepsilon)$, $A = A(D)$. Let $A_0 = \{f \text{ in } A;\ f(0) = f(\frac{1}{2})\}$. The algebra A_0 is the limit algebra of interest, its maximal ideal space is the degenerate Riemann surface obtained by starting with D and identifying the points z = 0 and z = ½ to a single point.

Proposition 2: $d(A_\varepsilon, A_0) = O(\varepsilon)$.

Proof: Again we construct T_ε in $L(A_\varepsilon, A_0)$ with $\|T_\varepsilon\| \, \|T_\varepsilon^{-1}\| = 1 + O(\varepsilon)$. Let R_ε be the map from A_ε to $\hat{A}_\varepsilon$ obtained by regarding functions on S_ε as defined on D_ε. Let Q_ε be the map from A_ε to A given by taking the Cauchy integral around the unit circle, i.e., if f is in A_ε and $|z| < 1$ then

$$(Q_\varepsilon f)(z) = \frac{1}{2\pi i} \int_{|\zeta|=1} f(\zeta) \frac{d\zeta}{\zeta - z}.$$

(If f is in A_ε then $f(z) = \frac{1}{2\pi i} \int_{\partial D_\varepsilon} f(\zeta) \frac{d\zeta}{\zeta - z}$, hence $Q_\varepsilon f$ is in A.)

Let P be the map from A to A_0 given by $(Pf)(z) = f(z) - 2(f(\tfrac{1}{2}) - f(0))z$. Let $T_\varepsilon = PQ_\varepsilon R_\varepsilon$. T_ε is a bounded map of A_ε into A_0. We now show that for all f in A_ε with $\|f\| \neq 0$,

$$\frac{\|T_\varepsilon f\|}{\|f\|} = 1 + O(\varepsilon). \tag{4.1}$$

First note that by the maximum principle on S_ε, for any f in A_ε, $\|R_\varepsilon f\| = \|f\|$ and $\|R_\varepsilon f\| = \sup \{|f(z)|;\ |z| = 1\}$. For such an f the Cauchy integral formula gives

$$R_\varepsilon f(z) - Q_\varepsilon R_\varepsilon f(z) = \frac{1}{2\pi i} \left(\int_{|\zeta|=\varepsilon} + \int_{|\zeta-\frac{1}{2}|=\varepsilon} \right) R_\varepsilon f(\zeta) \frac{d\zeta}{\zeta - z}.$$

Since $R_\varepsilon f(\zeta)$ is dominated by $\|f\|$, $(\zeta - z)$ is bounded away from zero for ζ on the path of integration and z on the unit circle, and the length of the path of integration is $O(\varepsilon)$, we can conclude that $|R_\varepsilon f(z) - Q_\varepsilon R_\varepsilon f(z)| = O(\varepsilon)$ for z with $|z| = 1$. Thus

$$\|f\| - \|Q_\varepsilon R_\varepsilon f\| = O(\varepsilon \|f\|) \tag{4.2}$$

for all f in A_ε. Let $g = Q_\varepsilon R_\varepsilon f$. To show that Pg and g are almost the same size, we will show that

$$|g(0) - g(\tfrac{1}{2})| = O(\varepsilon \|g\|). \tag{4.3}$$

By the Schwarz-Pick theorem

$$|g(0) - g(\varepsilon)| = O(\varepsilon \|g\|) \quad , \quad |(g(\tfrac{1}{2}) - g(\tfrac{1}{2} - \varepsilon)| = O(\varepsilon \|g\|) \tag{4.4}$$

for any g in A. $R_\varepsilon f$ is defined on D_ε, however the mapping of z to $\frac{1}{2} - \frac{\varepsilon^2}{z}$ which was used to identify $\{|z| = \varepsilon\}$ with $\{|z-\frac{1}{2}| = \varepsilon\}$ can be used to extend the $R_\varepsilon f$ so that it is analytic in $D \setminus [\{|z| \leq \varepsilon\} \cup \{|z - \frac{1}{2}| \leq 10\varepsilon^2\}]$. Hence, by the Cauchy integral formula, $g(\frac{1}{2} - \varepsilon) = (Q_\varepsilon R_\varepsilon f)(\frac{1}{2} - \varepsilon)$

$$= (R_\varepsilon f)(\tfrac{1}{2} - \varepsilon) - \frac{1}{2\pi i} \int_{|\zeta|=\varepsilon} R_\varepsilon f(\zeta) \frac{d\zeta}{\zeta - (\frac{1}{2} - \varepsilon)} - \frac{1}{2\pi i} \int_{|\zeta - \frac{1}{2}| = 10\varepsilon^2} R_\varepsilon f(\zeta) \frac{d\zeta}{\zeta - (\frac{1}{2} - \varepsilon)}.$$

Using trivial estimates for both integrals on the right we obtain

$$g(\tfrac{1}{2}-\epsilon) = (R_\epsilon f)(\tfrac{1}{2}-\epsilon) + O(\epsilon \|g\|).$$

Similarly

$$g(\epsilon) = (R_\epsilon f)(\epsilon) + O(\epsilon \|g\|).$$

However, since f was defined on S_ϵ,

$$(R_\epsilon f)(\epsilon) = (R_\epsilon f)(\tfrac{1}{2}-\epsilon).$$

Combining these three equalities with (4.4) yields (4.3). The definition of T_ϵ, (4.2) and (4.3) now give (4.1).

It follows from (4.1) that T_ϵ is a one to one continuous map onto a closed subspace of A_0, and that T_ϵ almost preserves norms of elements of A_ϵ. To complete the proof we must show that $T_\epsilon(A_\epsilon)$ is all of A_0. To do that it suffices to show that $T_\epsilon(A_\epsilon)$ has codimension one in A.

By regarding $D_\epsilon \cup \{|z| > 1\} \cup \{\infty\}$ as a subset of the Riemann sphere, we may regard S_ϵ as a subset of a compact Riemann surface of genus one $\hat{S}_\epsilon$. Consider the Weierstrass $\wp$ function of this surface with pole at ∞. $\wp$ is meromorphic on $\hat{S}_\epsilon$ and has a pole of exact order 2 at ∞. Hence $R_\epsilon \wp$ can be extended to a meromorphic function on $D_\epsilon \cup \{|z| > 1\} \cup \{\infty\}$ which is regular except for a pole of exact order 2 at ∞. Thus there is a polynomial P_2 of exact degree 2 such that $R_\epsilon \wp - P_2$ is analytic in $\{|z| > \frac{3}{4}\}$ and vanishes at ∞. But $R_\epsilon \wp - Q_\epsilon R_\epsilon \wp$ is also analytic in $\{|z| > \frac{3}{4}\}$ and vanishes at ∞. Hence the difference $P_2 - Q_\epsilon R_\epsilon P$ is analytic in $\{|z| > \frac{3}{4}\}$ and vanishes at ∞. However this last function is analytic in $\{|z| < 1\}$. Hence $Q_\epsilon R_\epsilon \wp = P_2$. Hence $T_\epsilon \wp$ is a polynomial of exact degree 2. Repeating this argument with $\wp'$ and with polynomials in $\wp$ and $\wp'$ we find that $T_\epsilon A_\epsilon$ contains polynomials of exact degree n for all integers $n \geq 2$. $T_\epsilon A_\epsilon$ also contains the constant functions. Hence the set X of functions of the form $\{a + cZ;\ a \in T_\epsilon A_\epsilon,\ c \in C\}$, where Z is the coordinate function on D, contains all polynomials. Since $T_\epsilon A_\epsilon$ is closed, $X = A$. Hence $T_\epsilon A_\epsilon$ has codimension one in A and the proof is complete.

5. Pinching a Boundary. This construction shows what can happen if we start with an S in S(g,1) for some positive g and pinch to a point a curve which is homotopic to ∂S. From a geometric point of view one might expect the limiting object to be that obtained from S_g, a compact surface of genus g, and $S_0 = \{|z| \leq 1\}$ by identifying a point p of S_g with a point q of S_0. However S_g carries no non-constant analytic functions and hence does not appear in the limiting object. What does appear is the ghost of S_g, namely g linear conditions at the point q which the functions in the limiting algebra A_0 must satisfy.

Pick and fix a positive integer g, and S a compact Riemann surface of genus g. Pick and fix p_0 a non-Weierstrass point of S. Pick and fix $B_{g+1}, \cdots, B_{2g+1}$ meromorphic functions on S such that B_k has a pole of exact order k at p_0 and no other poles on S. Let ζ be the local uniformizer at p_0 given by $\zeta = (B_{g+1})^{-\frac{1}{g+1}}$. Thus $\zeta(p) = 0$, and near p_0, $B_{g+1}(p) = \zeta(p)^{-(g+1)}$ and for $k = 2,\cdots, g$ (after changing B_{g+k} by a factor)

$$B_{g+k} = \frac{1}{\zeta^{g+k}} + \sum_{-g-k+1}^{\infty} \alpha_{kj}\zeta^j .$$

For ϵ sufficiently small, $\{|\zeta(p)| = \epsilon\}$ is a simple closed curve about p_0. For such ϵ let $S_\epsilon = S\setminus\{|\zeta(p)| < \epsilon\}$. Thus $S_\epsilon \in S(g,1)$. Let $A_\epsilon = A(S_\epsilon)$. Let $A_0 = \{f;\ f$ continuous in $|z| \le 1$, f analytic in $|z| < 1$, $f'(0) = \cdots = f^{(g)}(0) = 1\}$.

<u>Proposition 3</u>: $\lim_{\epsilon\to 0} d(A_\epsilon, A_0) = 0$.

<u>Proof</u>: Again we construct $T_\epsilon \in L(A_0,A_\epsilon)$ with $\lim_{\epsilon\to 0} \|T_\epsilon\|\,\|T_\epsilon^{-1}\| = 1$. For any f in $A = A(\{|z| \le 1\})$

$$f(z) = f_1(z^{g+1}) + zf_2(z^{g+1}) + \cdots + z^g f_{g+1}(z^{g+1})$$

where the f_i are elements of A which are uniquely determined by this formula and which depend continuously on f. Hence for any f in A_0

$$f(z) = f_1(z^{g+1}) + z^{g+2}f_2(z^{g+1}) + \cdots + z^{2g+1}f_{g+1}(z^{g+1}) \tag{5.1}$$

where the f_i are uniquely determined by (5.1). For some constant B, $\|f_k\| \le B\|f\|$. Pick ϵ small, define $TF = T_\epsilon f$ by

$$Tf = f_1(T(z^{g+1})) + T(z^{g+2})f_2(T(z^{g+1})) + \cdots + T(z^{2g+1})f_{g+1}(T(z^{g+1}))$$

where

$$T(z^{g+k}) = \epsilon^{g+k}B_{g+k} \qquad k = 1, \cdots, g.$$

It follows from the uniqueness and continuity of the decomposition (5.1) that T is a well defined continuous linear map of A_0 into A_ϵ and that $T(1) = 1$. On ∂S_ϵ

$$T(z^{g+k}) = \epsilon^{g+k}B_{g+k} = \epsilon^{g+1}\cdot\frac{1}{\zeta^{g+1}} \qquad (k = 1)$$

$$= \frac{\epsilon^{g+k}}{\zeta^{g+k}} + \epsilon^{g+k}\sum_{-g-k+1}^{\infty}\alpha_{kj}\zeta^j \qquad (1 < k < g).$$

Let $L_k = T(z^{g+k}) - T(z^{g+1})^{(g+k)/(g+1)}$. By the previous equation, $L_1 = 0$ and, for $2 \le k \le g+1$,

$$L_k = \epsilon^{g+k}\zeta^{-g-k+1}h_k(\zeta) = \epsilon\cdot\left(\frac{\epsilon}{\zeta}\right)^{g+k-1}h_k(\zeta)$$

where $h_k(\zeta)$ is bounded in a neighborhood of p_0. Hence there is a constant B' that is independent of ϵ and k such that

$$\sup_{\partial S_\epsilon} |L_k| \leq B' \epsilon.$$

On ∂S_ϵ,

$$Tf = f_1(T(z^{g+1})) + \sum_{k=2}^{g+1} T(z^{g+k})^{(g+k)/(g+1)} f_k(T(z^{g+1}))$$

$$+ \sum L_k f_k(T(z^{g+1}))$$

$$= f(T(z^{g+1})^{\frac{1}{g+1}}) + \sum L_k f_k(T(z^{g+1}))$$

$$= f(\epsilon/\zeta) + \sum L_k f_k(T(z^{g+1})).$$

Hence

$$|Tf(\zeta) - f(\epsilon/\zeta)| \leq \sum \|L_k\| \, \|f_k(T(z^{g+1}))\|$$

$$\leq \sum B' \epsilon \, \|f_k\|$$

$$\leq B'' \epsilon \, \|f\|.$$

From this estimate it follows easily that there is an ϵ' which can be made arbitrarily small if ϵ is arbitrarily small so that for all f in A_0

(5.2) $$1 - \epsilon' \leq \frac{\|T_\epsilon f\|}{\|f\|} \leq 1 + \epsilon'.$$

Hence T is one to one and has closed range. Now, if we show $T(A_0)$ is dense in A_ϵ it will follow that $T_\epsilon \in L(A_0, A_\epsilon)$. It will then follow from (5.2) that $\|T_\epsilon\| \, \|T_\epsilon^{-1}\| \to 1$ as $\epsilon \to 0$.

From the definition of T_ϵ and the fact that p_0 is not a Weierstrass point we see that $T_\epsilon(A_0)$ contains the restriction to S_ϵ of every function meromorphic on S and regular on $S \setminus \{p\}$. Royden has shown [10, Th. 10] that this class of functions can be used to give arbitrarily good uniform approximation on S_ϵ to any function in $A_{\epsilon'}$ for any $\epsilon' < \epsilon$. By a theorem of Alling [1, Theorem 7.5] the set of functions $\bigcup_{\epsilon' < \epsilon} A_{\epsilon'}$ is dense in $A_{\epsilon'}$. Hence $T_\epsilon(A_0)$ is dense in A_ϵ and the proof is complete.

<u>Note</u>: By filling in the details of the estimates one finds $d(A_\epsilon, A_0) = O(\epsilon)$.

6. <u>Consequences of the Constructions</u>. A. The main goal of these constructions was to establish the following. Suppose g, k are integers, $g \geq 0$, $k \geq 1$, $(g,k) \neq (0,1)$.

<u>Proposition 4</u>: The metric space $(R(g,k), d)$ is not complete.

Proof: Suppose first that $k \geq 2$. Let ϵ_n be positive numbers tending to zero. Pick S_1 in $S(g,k-1)$ and S_2 in $S(0,1)$ and let S_n be the surface S_{ϵ_n} constructed from S_1 and S_2 using the construction of Section 3. By Proposition 1, $\{S_n\}_{n=3}^{\infty}$ is a Cauchy sequence in $(R(g,k),d)$. We now wish to show that the sequence $\{S_n\}$ has no limit point in $(R(g,k),d)$. This can be done by using a general result about convergence in the Banach-Mazur metric. For S in S let $m(S) = \inf\{\ell(\gamma)\ ;\ \gamma$ a simple closed curve on S which is not homologically trivial$\}$. If $\{S_n\}$ are elements of S and $S_n \to S$ in the d-metric, then $\lim m(S_n) = m(S)$ ([8],[9]). For our $\{S_n\}$, $m(S_n)$ tends to zero. If S is in S, then $m(S) > 0$. Hence $\{S_n\}$ has no limit point in $(R(g,k),d)$. (This could have been shown directly by adapting the arguments of [7].) If $k = 2$, then we argue the same way except that we use the construction of Section 4 and Proposition 2.

Corollary: On $R(g,k)$, the ratio of the Teichmuller distance between points to the d distance is unbounded.

Proof: The Teichmuller metric on $R(g,k)$ is known to be complete.

B. For a function algebra A, let ∂_A denote the Shilov boundary of A and M_A the maximal ideal space of A. Call A nice if every point of ∂_A is a strong peak point, i.e. given x in ∂_A there is a in A so that $x(a) = 1$ and $|y(a)| < 1$ for all y in ∂_A, $y \neq x$. Small changes of A can change M_A but cannot change ∂_A.

Proposition 5: A) There is a universal constant $c > 0$ such that if A and B are nice function algebras and $d(A,B) < c$ then ∂_A and ∂_B are homeomorphic.

B) There is a nice function algebra A_0 such that for all $\epsilon > 0$ there is a nice function algebra A_ϵ with $d(A_0,A_\epsilon) < \epsilon$ and the first homology group of M_{A_ϵ} is different from that of M_{A_0}. In particular M_{A_ϵ} and M_{A_0} are not homeomorphic.

Proof: Part A is Theorem A of [6]. Both Section 3 and Section 5 provide examples of A_0 which satisfy B.

C. If A_1 and A_2 are Banach spaces, we will write $A_1 \sim A_2$ to denote that A_1 and A_2 are isomorphic (i.e. $d(A_1,A_2) < \infty$).

Proposition 6: Suppose $S_1, S_2 \in S$. Then $A(S_1) \sim A(S_2)$.

Although a proof of this result has never been published, various special cases are well known to a number of people and the result itself can be regarded as known in the folklore.

For the remainder of this section let A denote the disk algebra. $A = A(D)$. The proposition follows immediately from the following three elementary lemmas, the previous constructions, and the results of [7].

Lemma 1: Let $A_1, \cdots, A_n$ be n copies of A, then $A \sim A_1 + \cdots + A_n$.

Proof: It suffices to prove this for n = 2. If f is in A, then the decomposition of f as the sum of an odd function plus an even function allows us to write $f(Z) = f_1(Z^2) + Zf_2(Z^2)$. It is elementary to verify that $f \to (f_1, f_2)$ is an isomorphism of A and $A_1 + A_2$.

Lemma 2: Let n be a positive integer and let $A_{(n)} = \{f \in A \; ; \; f'(0) = f''(0) = \cdots = f^{(n)}(0) = 0\}$, then $A \sim A_{(n)}$.

Proof: $f \to f(0) + Z^n(f - f(0))$ is the required isomorphism.

Let $S_i \in S$, $i = 1,2$. Pick $p_i \in \text{int } S_i$, $i = 1,2$. Let $\hat{S}$ be the degenerate Riemann surface formed by identifying p_1 and p_2. Let $A(\hat{S})$ be the algebra of all continuous boundary value analytic functions on $\hat{S}$, equivalently $A(\hat{S})$ consists of pairs (f_1, f_2) with $f_i \in A(S_i)$ and $f_1(p_1) = f_2(p_2)$.

Lemma 3: $A(S_1) + A(S_2) \sim A(\hat{S})$.

Proof: Pick $g \in A(S_2)$ such that $g(p_2) = 0$ and g has no other zeros on S_1. The map $(f_1, f_2) \to (f_1, f_1(p_1) + gf_2)$ is the required isomorphism of $A(S_1) + A(S_2)$ and $A(\hat{S})$. We again omit the elementary verification.

Proof of Proposition 6: Pick $S \in S(g,k)$. It suffices to show that A(S) is isomorphic to the disk algebra A = A(D). By the first lemma, $A \sim A_1 \oplus \cdots \oplus A_k$, with A_i, $i = 1,2, \cdots, k$ isomorphic copies of A. Now apply the second lemma to the first summand to obtain $A \sim A_{(g)} \oplus A_2 \oplus \cdots \oplus A_k$. By proposition 5, $A_{(g)} \sim A(R)$ for some $R \in S(g,1)$. Thus $A \sim A(R) \oplus A_2 \oplus \cdots \oplus A_k$. By the third lemma, $A(R) \oplus A_2 \sim A(\hat{R})$ where $\hat{R}$ is the degenerate Riemann surface obtained by identifying a point p in R with the origin of the unit disk. Thus $A \sim A(\hat{R}) \oplus A_3 \oplus \cdots \oplus A_k$. By Proposition 3 $A(\hat{R}) \sim A(R')$ for some R' in S(g,2). Thus $A \sim A(R') \oplus A3 \oplus \cdots \oplus A_k$. Continuing in this way we find $A \sim A(R'')$ for some R'' in S(g,k). By Theorem 2 of [7], $A(R'') \sim A(S)$. Thus $A \sim A(S)$ and the proof is complete.

These arguments could actually yield a bit more. We will only sketch the extension. Let $\hat{S}$ be the class of not necessarily connected degenerate finite bordered Riemann surfaces. That is, an element of $\hat{S}$ is formed by starting with a finite union of elements of S and making finitely many identifications of finite sets of interior points. For S in $\hat{S}$, let $A(\hat{S})$ be the algebra of all continuous boundary value analytic functions on $\hat{S}$.

Proposition 6': Suppose that for i = 1,2 , $\hat{S}_i$ is in $\hat{S}$ and A_i is a subalgebra if $A(\hat{S}_i)$ of finite codimension, then $A_1 \sim A_2$.

Proof Outline: It suffices to show A_1 is isomorphic to the disk algebra A. Suppose

$\hat{S}_1$ was constructed from $S_1, S_2, \cdots, S_n$. By the previous proposition and Lemma 1, $A \sim A(S_1) \oplus \cdots \oplus A(S_n)$. By application of a straightforward extension of Lemma 3, this implies $A \sim A(\hat{S}_1)$. Gamelin has shown [2] that any subalgebra of $A(\hat{S}_1)$ of finite codimension is obtained by making further identification of interior points of $\hat{S}_1$ and by taking kernels of point derivations supported at interior points of $\hat{S}_1$. By straightforward extensions of Lemma 3 and Lemma 2, neither of these processes changes the isomorphism class. The proof outline is complete.

Finally, most of the arguments just used are valid mutatis mutandis for algebras of the form $H^\infty(S)$, all bounded analytic functions on S.

Proposition 6": If $\hat{S}_1$, $\hat{S}_2$ are in $\hat{\mathcal{S}}$ then $H^\infty(\hat{S}_1) \sim H^\infty(\hat{S}_2)$.

7. Questions. There are many interesting open questions related to the results of the previous section. We mention a few.

First, what is the Cauchy completion of $(R(g,k),d)$? It can be shown [8], [9] that any Cauchy sequence in $(R(g,k),d)$ has a limit which is a subalgebra of an algebra of the form $A(S)$, $S \in \mathcal{S}$. Must this subalgebra be one that can be constructed by the three types of degenerative processes described in Sections 3, 4, and 5? In particular is the limit a subalgebra with finite codimension in some $A(S)$, $S \in \mathcal{S}$?

Second, the constructions of Sections 3, 4, and 5 can be regarded as deformations of function algebras induced by deformations of the conformal structure of the maximal ideal space. The unit disk does not admit non-trivial infinitesimal deformations of its conformal structure. This suggests the following. Let A be the disk algebra.

Conjecture: There is a positive constant c such that if B is a nice function algebra and $d(A,B) < c$ then $A = B$ (i.e. there is an isometric algebra isomorphism of A and B.)

The analog of this conjecture is true for some algebras. For example, let T be the unit circle and C(T) the algebra of all continuous complex valued functions on T.

Proposition 7: There is a positive constant c such that if B is a nice function algebra and $d(C(T),B) < c$ then $C(T) = B$.

Proof: By Theorem A of [6], if c is small then we may conclude that ∂_B is homeomorphic to T and thus regard B as a subalgebra of C(T). If $d(C(T),B)$ is small then there is an R in $L(C(T),B)$ with $\|R\| \, \|R^{-1}\| \sim 1$. Using Theorem A again, $R = L + K$, where L is an automorphism of C(T) and K has small norm. Hence $B = R(C(T))$ is all of C(T). The proof is complete.

Finally, we have seen that the class of nice function algebras B with $A \sim B$ contains many different algebras. Exactly what are the elements of this class? In particular is there a B such that $A \sim B$ and M_B contains no analytic structure? Is there a Riemann surface S of infinite connectivity with $A \sim A(S)$? For a discussion of the general problem of isomorphism of algebras of analytic functions see [5].

REFERENCES

1. N. Alling, Extensions of Meromorphic Function Rings over Non-compact Riemann Surfaces II, Math. Zeitschr. 93(1966), 345-394.
2. T. W. Gamelin, Embedding Riemann Surfaces in Maximal Ideal Spaces, J. Functional Anal., Vol. 2(1968), 123-146.
3. H. J. Landau and R. Osserman, On Analytic Mappings of Riemann Surfaces, J. Analyse Math., 7(1959), 249-279.
4. A. Lebowitz, On the Degeneration of Riemann Surfaces, Advances in the Theory of Riemann Surfaces, Princeton University Press, 1971.
5. A. Pelczynski, Banach Spaces of Analytic Functions, To Appear, Regional Conference Series in Mathematics AMS, Providence, RI.
6. R. Rochberg, Almost Isometries of Banach Spaces and Moduli of Planar Domains, Pac. J. Math., 49(1973), 445-466.
7. _________, Almost Isometries of Banach Spaces and Moduli of Riemann Surfaces, II, Duke Math. J., 42(1975), 167-182.
8. _________, Algebras of Analytic Functions on Degenerating Riemann Surfaces, Bull. Amer. Math. Soc. 81(1975), 202-204.
9. _________, Function Algebras Obtained as Limits in the Banach-Mazur Metric (in preparation).
10. H. Royden, Function Theory on Compact Riemann Surfaces, Journal D'Analyse Math. 43(1967), 295-327.
11. L. Ahlfors and L. Sario, Riemann Surfaces, Princeton University Press, Princeton, N.J., 1960.

OPERATOR THEORY IN HARMONIC ANALYSIS*

Bernard Russo**
University of California
Irvine, CA 92717/USA

Abstract. The purpose of this paper is to give an illustration of how operator theoretic results in Hilbert space can be applied to obtain results in classical and abstract harmonic analysis. An inequality for integral operators will be used to give new proofs for the classical Hausdorff-Young Theorems on the unit circle (Fourier coefficients) and on the real line (Fourier integral). These proofs were discovered in the course of the author's investigations into Hausdorff-Young phenomena on non-commutative non-compact groups. The proof for the unit circle is short and will be given in §1. The proof for the real line is more involved and will be given in §2. In §3 a brief summary of the evolution of abstract harmonic analysis is given which includes the statement of the Hausdorff-Young Theorem for unimodular groups. In §4 the question of sharpness in the Hausdorff-Young Theorem in various settings is discussed and recent results of W. Beckner, J. Fournier and the author are described.

1. Fourier Coefficients. Let f be a measurable periodic function on $\mathbb{R}$ of period 2π with Fourier coefficients $\hat{f}(n) = \frac{1}{2\pi}\int_0^{2\pi} f(x)e^{-inx}dx$, $n \in \mathbb{Z}$. The classical Riesz-Fischer Theorem [I,II] (1907) asserts that

$$(1.1) \qquad \sum_{n=-\infty}^{\infty} |\hat{f}(n)|^2 = \frac{1}{2\pi}\int_0^{2\pi} |f(x)|^2 dx.$$

More precisely, the map $f \to \hat{f}$ is an isometry of $L^2(\mathbb{T})$ onto $L^2(\mathbb{Z})$ where $\mathbb{T}$ denotes the unit circle with normalized Haar measure and 2π periodic functions are identified with functions on $\mathbb{T}$. We rewrite (1.1) as

$$(1.2) \qquad \|\hat{f}\|_{L^2(\mathbb{Z})} = \|f\|_{L^2(\mathbb{T})}, \quad f \in L^2(\mathbb{T}).$$

The Riesz-Fischer Theorem was first extended by Young in 1912-13 [III,IV] as follows. Let m be an integer, $m \geq 2$, and let $p = 2m/(2m-1)$. Then the conjugate index p', defined by $\frac{1}{p} + \frac{1}{p'} = 1$ is equal to 2m and is thus an even integer.

---

*Lecture at the Regional Conference on Banach spaces of analytic functions, Kent State University, July, 1976, supported by National Science Foundation.

**Research supported by National Science Foundation grant MCS-76-07219.

This sequence $p = 2m/(2m-1)$ has played an important role in the subject and will play a key role in this paper when we discuss Fourier integrals. By exploiting the relation between Fourier coefficients and convolution and by using the inequality now known as Young's inequality, Young showed that

$$(1.3) \qquad \left(\sum_{n=-\infty}^{\infty} |\hat{f}(n)|^{p'}\right)^{1/p'} \le \left(\frac{1}{2\pi}\int_0^{2\pi} |f(x)|^p dx\right)^{1/p}$$

$$\text{for } p = 2m/(2m-1)\ , \ m \ge 2,$$

i.e.,

$$(1.4) \qquad \|\hat{f}\|_{L^{p'}(\mathbb{Z})} \le \|f\|_{L^p(T)}\ , \quad f \in L^p(\mathbb{T})$$

$$\text{for } p = 2m/(2m-1)\ , \ m \ge 2.$$

The inequality (1.4) was extended to all values of p in $(1,2)$ by Hausdorff in 1923 [V] and the inequality became known as the Hausdorff-Young Theorem. The idea of deriving the Hausdorff-Young Theorem for all $p \in (1,2)$ by using a convexity theorem together with its validity for $p = 1$ and $p = 2$ goes back to M. Riesz [VI] (1926).

We shall now show how for each fixed $p \in (1,2)$ (1.4) is a special case of a theorem due to the author on integral operators. To state this theorem we recall first that if T is a bounded linear operator on an arbitrary Hilbert space $\mathcal{H}$, norms $\|T\|_p$ are defined by $\|T\|_p = [\text{trace } (T^*T)^{p/2}]^{1/p}$ if $1 \le p < \infty$; and second, if $k \in L^2(X \times X)$ where X is a σ-finite measure space, then $k^*(x,y) = \overline{k(y,x)}$ and norms $\|k\|_{p,q}$ are defined by $\|k\|_{p,q} = \left[\int_X \left[\int_X |k(x,y)|^p dx\right]^{q/p} dy\right]^{1/q}$ if $p,\ q \in [1,\infty)$. The following is called the Hausdorff-Young Theorem for integral operators [10,12]: let $k \in L^2(X \times X)$ and let T_k be the integral operator with kernel k: $(T_k f)(x) = \int_X k(x,y)f(y)dy$, $f \in L^2(X)$, $x \in X$. If $p \in (1,2)$, then with $\frac{1}{p} + \frac{1}{p'} = 1$,

$$(1.5) \qquad \|T_k\|_{p'} \le (\|k\|_{p,p'} \|k^*\|_{p,p'})^{1/2}.$$

To obtain (1.4) for any $p \in (1,2)$ from (1.5) let X be the unit circle with measure $dx/2\pi$ and for an integrable function f with Fourier coefficients $\{\hat{f}(n)\}_{-\infty}^{\infty}$ let k_f be the function on $X \times X$ given by $k_f(x,y) = f(x-y)$. Then T_{k_f} is just the convolution operator $g \to f*g$ and thanks to the relation $(f*g)^\wedge = \hat{f}\cdot\hat{g}$ and (1.2) the operator T_{k_f} is unitarily equivalent to the operator $M_{\hat{f}}$ of multiplication by $\hat{f}$ on $L^2(\mathbb{Z})$. It follows that the singular values of $M_{\hat{f}}$ (i.e., the eigenvalues of $((M_{\hat{f}})^* M_{\hat{f}})^{1/2}$) are $\{|\hat{f}(n)|\}$ and thus $\|T_{k_f}\|_r = \|M_{\hat{f}}\|_r = \left(\sum_{n\in\mathbb{Z}} |\hat{f}(n)|^r\right)^{1/r} = \|\hat{f}\|_{L^r(\mathbb{Z})}$ for all indices $r \in [1,\infty)$. On the other hand $\|k_f\|_{p,r} = \|f\|_{L^p(T)} = \|(k_f)^*\|_{p,r}$ for all indices r (here we used the compactness of $\mathbb{T}$). Thus $\|f\|_{L^p(\mathbb{T})} = (\|k_f\|_{p,p'}\|(k_f)^*\|_{p,p'})^{1/2} \geq \|T_{k_f}\|_{p'} = \|\hat{f}\|_{L^{p'}(\mathbb{Z})}$ as required.

We close this section with the remark that by a standard elementary duality argument, (1.4) for any fixed p, implies its counterpart on $\mathbb{Z}$ i.e., for any $p \in (1,2)$,

$$(1.6) \qquad \|\hat{\phi}\|_{L^{p'}(\mathbb{T})} \leq \|\phi\|_{L^p(\mathbb{Z})}, \quad \phi \in L^p(\mathbb{Z}) \cap L^1(\mathbb{Z})$$

where $\frac{1}{p} + \frac{1}{p'} = 1$ and

$$(1.7) \qquad \hat{\phi}(\lambda) = \sum_{m=-\infty}^{\infty} \phi(m)\lambda^{-m}, \quad \lambda \in \mathbb{T}.$$

The inequality (1.6) will be used in our discussion of Fourier integrals.

2. Fourier Integrals. Let f be an integrable function on $\mathbb{R}$. The Fourier transform of f can be defined by $\hat{f}(y) = (2\pi)^{-\frac{1}{2}} \int_{\mathbb{R}} e^{-ixy} f(x)dx$ for $y \in \mathbb{R}$ and the analog of the Riesz-Fischer theorem, due to Plancherel in 1910 [VII] asserts that

$$(2.1) \qquad \int_{\mathbb{R}} |\hat{f}(y)|^2 dy = \int_{\mathbb{R}} |f(x)|^2 dx, \quad f \in L^1(\mathbb{R}) \cap L^2(\mathbb{R}).$$

This can be written

$$(2.2) \qquad \|\hat{f}\|_{L^2(\mathbb{R})} = \|f\|_{L^2(\mathbb{R})}, \quad f \in L^1(\mathbb{R}) \cap L^2(\mathbb{R}).$$

By the use of Riesz convexity one obtains the Hausdorff-Young Theorem for the real line: for $p \in (1,2)$ and $\frac{1}{p} + \frac{1}{p'} = 1$,

$$(2.3) \qquad \left((2\pi)^{-\frac{1}{2}} \int_{\mathbb{R}} |\hat{f}(y)|^{p'} dy\right)^{1/p'} \leq \left((2\pi)^{-\frac{1}{2}} \int_{\mathbb{R}} |f(x)|^p dx\right)^{1/p}, \quad f \in L^1(\mathbb{R}) \cap (L^p(\mathbb{R}).$$

This can be written

$$(2.4) \qquad \|\hat{f}\|_{L^{p'}(\mathbb{R})} \leq \|f\|_{L^p(\mathbb{R})}, \quad f \in L^1(\mathbb{R}) \cap L^p(\mathbb{R}),$$

where the constant $(2\pi)^{-\frac{1}{2}}$ must be considered as part of the measure.

The inequality (2.4) was proved first by Titchmarsh in 1924 [VIII]. The objective of this section is to give another proof of (2.4) for values of p of the form $p = 2m/(2m-1)$. The proof given in 1 for the corresponding result on the unit circle and for all $p \in (1,2)$ will not work here because of the non-compactness of $\mathbb{R}$ and the non-compactness of convolution operators on $L^2(\mathbb{R})$. We overcome the non-compactness of $\mathbb{R}$ by the facts that we know the analogous result on $\mathbb{Z}$ (formula

(1.6)) and that $\mathbb{R}$ is a compact extension of $\mathbb{Z}$, i.e., $\mathbb{R}/\mathbb{Z}$ is compact. In place of the non-compactness of convolution operators we shall derive by elementary methods a formula for the norm $\|\hat{f}\|_{L^{p'}(\mathbb{R})}$ which amounts to expressing the operator of convolution by f as a direct integral of compact operators (namely, convolution operators on $L^2(\mathbb{T})$).

We proceed to the proof of (2.4). We first recall that the map $k \to T_k$ from kernels to operators is well behaved and in particular the kernal h of $(T_k)^*T_k$ is given by $h(x,y) = \int_X \overline{k(z,x)}k(z,y)dz$. Now let $s\colon \mathbb{T} \to \mathbb{R}$ be defined by $e^{is(\alpha)} = \alpha$ and $s(\alpha) \in [0,2\pi)$. Let $d\alpha$ denote normalized Haar measure on $\mathbb{T}$, i.e., $d\alpha = \frac{ds(\alpha)}{2\pi}$. For any $\varphi \in L^1(\mathbb{R})$ we have the integration formula

$$\int_{\mathbb{R}} \varphi(x)dx = 2\pi \sum_{n=-\infty}^{\infty} \int_{\mathbb{T}} \varphi(s(\alpha)+2\pi n)d\alpha \ .$$

For any function $\varphi \in L^1(\mathbb{R})$ and pair $\alpha,\beta \in \mathbb{T}$ define a function $f_{\varphi,\alpha,\beta} \in L^1(\mathbb{Z})$ by $f_{\varphi,\alpha,\beta}(n) = \varphi(2\pi n+s(\alpha)-s(\beta))$. Also for each $\lambda \in \mathbb{T}$ define a kernel $k(\varphi,\lambda)$ on $\mathbb{T} \times \mathbb{T}$ by $k(\varphi,\lambda;\alpha,\beta) = (f_{\varphi,\alpha,\beta})^\wedge(\lambda)$ (Fourier transform on $\mathbb{Z}$ as in (1.7)). Suppose we prove

$$\|\hat{\varphi}\|^{2m}_{L^{2m}(\mathbb{R})} = (2\pi)^{\frac{1}{2}(2m-1)} \int_{\mathbb{T}} \|T_{k(\varphi,\lambda)}\|^{2m}_{2m} d\lambda \tag{2.5}$$

for m = 1,2,3,... where the norms inside the integral are defined by the trace as in §1. For notation's sake let p = 2m/(2m-1) in the rest of this proof. Then by (1.5) $\|T_{k(\varphi,\lambda)}\|_{2m} = \|T_{k(\varphi,\lambda)}\|_{p'} \le (\|k(\varphi,\lambda)\|_{p,p'}\|k(\varphi,\lambda)^*\|_{p,p'})^{\frac{1}{2}}$ and thus by (2.5) we have

$$\|\hat{\varphi}\|^{p'}_{L^{p'}(\mathbb{R})} \le (2\pi)^{\frac{1}{2}(2m-1)} \int_{\mathbb{T}} (\|k(\varphi,\lambda)\|_{p,p'}\|k(\varphi,\lambda)^*\|_{p,p'})^{p'/2} d\lambda$$

$$\le (2\pi)^{\frac{1}{2}(2m-1)} \left(\int_{\mathbb{T}} \|k(\varphi,\lambda)\|^{p}_{p,p'} d\lambda\right)^{\frac{1}{2}} \left(\int_{\mathbb{T}} \|k(\varphi,\lambda)^*\|^{p'}_{p,p'} d\lambda\right)^{\frac{1}{2}} .$$

Each of these integrals is dominated by $(2\pi)^{-p'/2p}\|\varphi\|^{p'}_{L^p(\mathbb{R})}$.

For example, $\int_{\mathbb{T}} \|k(\varphi,\lambda)\|^{p'}_{p,p'} d\lambda$

$$= \iint\left(\int |k(\varphi,\lambda;\alpha,\beta)|^p d\alpha\right)^{p'/p} d\beta d\lambda = \iint\left(\int |(f_{\varphi,\alpha,\beta})^\wedge(\lambda)|^p d\alpha\right)^{p'/p} d\beta d\lambda$$

$$\le \int\left(\int\left(\int |(f_{\varphi,\alpha,\beta})^\wedge(\lambda)|^{p'} d\lambda\right)^{p/p'} d\alpha\right)^{p'/p} d\beta \quad \text{(by Minkowski's integral inequality)}$$

$$= \int\left(\int \|(f_{\varphi,\alpha,\beta})^\wedge\|^p_{L^{p'}(\mathbb{T})} d\alpha\right)^{p'/p} d\beta$$

$$\le \int\left(\int \|f_{\varphi,\alpha,\beta}\|^p_{L^p(\mathbb{Z})} d\alpha\right)^{p'/p} d\beta \quad \text{(by equation (1.6))}$$

$$= \int\Big(\int \sum_n |\varphi(2\pi n+s(\alpha)-s(\beta)|^p d\alpha\Big)^{p'/p} d\beta$$

$$= \int\Big((2\pi)^{-1}\int_{\mathbb{R}} |\varphi(y-s(\beta)|^p dy\Big)^{p'/p} d\beta$$

$= (2\pi)^{-p'/2p}\|\varphi\|^{p'}_{L^p(\mathbb{R})}$ and similarly for the second integral. Therefore $\|\hat\varphi\|^{p'}_{L^{p'}(\mathbb{R})} \leq (2\pi)^{\frac{1}{2}(2m-1-p'/p)}\|\varphi\|^{p'}_{L^p(\mathbb{R})} = \|\varphi\|^{p'}_{L^p(\mathbb{R})}$ and we have proved (2.4) assuming (2.5). To prove (2.5) consider first the case m = 1. Then

$$\int_{\mathbf{T}} \|T_{k(\varphi,\lambda)}\|_2^2 d\lambda$$

$$= \iiint |k(\varphi,\lambda;\alpha,\beta)|^2 d\alpha d\beta d\lambda = \iint \|(f_{\varphi,\alpha,\beta})^{\wedge}\|^2_{L^2(\mathbf{T})} d\alpha d\beta$$

$$= \iint \|f_{\varphi,\alpha,\beta}\|^2_{L^2(\mathbf{Z})} d\alpha d\beta = \iint \sum_n |\varphi(2\pi n+s(\alpha)-s(\beta)|^2 d\alpha d\beta$$

$$= \int_{\mathbf{T}} (2\pi)^{-1}\int_{\mathbb{R}} |\varphi(y-s(\beta)|^2 dy d\beta = (2\pi)^{-\frac{1}{2}}\|\varphi\|^2_{L^2(\mathbb{R})} = (2\pi)^{-\frac{1}{2}}\|\hat\varphi\|^2_{L^2(\mathbb{R})}.$$

For m > 1 one proceeds carefully by iteration as follows. Let $\psi = \varphi^* * \varphi$. Then $\|\hat\varphi\|^4_{L^4(\mathbb{R})} = \|\hat\psi\|^2_{L^2(\mathbb{R})} = (2\pi)^{\frac{1}{2}}\int_{\mathbf{T}} \|T_{k(\psi,\lambda)}\|_2^2 d\lambda = (2\pi)^{3/2}\int_{\mathbf{T}} \|T_{k(\varphi,\lambda)}\|_4^4 d\lambda$ where the last step follows from

$$T_{k(\psi,\lambda)} = (2\pi)^{\frac{1}{2}}(T_{k(\varphi,\lambda)})^* T_{k(\varphi,\lambda)}. \tag{2.6}$$

To prove (2.6) one works with the kernels on both sides. Letting h denote the kernel of $(T_{k(\varphi,\lambda)})^* T_{k(\varphi,\lambda)}$ we have $h(\alpha,\beta) =$

$$\int_{\mathbf{T}} \sum_n \sum_n \overline{\varphi(2\pi n+s(\gamma)-s(\alpha))}\,\lambda^n \varphi(2\pi m+s(\gamma)-s(\beta))\lambda^{-m} d\gamma .$$

On the other hand, $k(\psi,\lambda;\alpha,\beta) = (f_{\psi,\alpha,\beta})^{\wedge}(\lambda) =$

$$\sum_k f_{\psi,\alpha,\beta}(k)\lambda^{-k} = \sum_k \psi(2\pi k+s(\alpha)-s(\beta))\lambda^{-k}$$

$$= \sum_k (2\pi)^{-\frac{1}{2}}\int_{\mathbb{R}} \overline{\varphi(-2\pi k-s(\alpha)+s(\beta)+y)}\,\varphi(y)dy\,\lambda^{-k}$$

$$= \sum_k (2\pi)^{-\frac{1}{2}}\int_{\mathbb{R}} \overline{\varphi(-2\pi k-s(\alpha)+y)}\,\varphi(y-s(\beta))dy\,\lambda^{-k}$$

$$= (2\pi)^{\frac{1}{2}}\sum_k\sum_{\ell}\int_{\mathbf{T}} \varphi(-2\pi k-s(\alpha)+\ell+s(\gamma)-s(\beta))\lambda^{-k}d\gamma$$

$= (2\pi)^{\frac{1}{2}}h(\alpha,\beta)$ (by setting $k = \ell - n$). This completes the proof of (2.5) for m = 2. The proofs for m = 3,4,... are similar. For example, for m = 3 let $\psi = \varphi * \varphi^* * \varphi$. Then the step analogous to (2.6) is

$$T_{k(\psi,\lambda)} = (2\pi) T_{k(\varphi,\lambda)} (T_{k(\varphi,\lambda)})^* T_{k(\varphi,\lambda)}.$$

3. Abstract Harmonic Analysis. The subject now called abstract harmonic analysis was a natural outgrowth of the work in the 1920's and 1930's of Weyl, Peter, Bochner, and Wiener on convolution operators and of Haar, Pontryagin and van Kampen on topological groups. It's goal was to extend results on Fourier series and integrals to locally compact Abelian groups and to compact non-Abelian groups. Great impetus was given to the subject around 1940 with the appearance of a pioneering monograph by Weil [14] and the development of the theory of commutative Banach algebras by Gelfand.

As an illustration suppose G is any locally compact Abelian group and let $\hat{G}$ denote the character group dual to G. For each Haar integrable function f on G the Fourier transform of f is the function $\hat{f}$ defined on $\hat{G}$ by

$$\hat{f}(\gamma) = \int_G f(x)\overline{\gamma(x)}dx \ , \ \gamma \in \hat{G}. \tag{3.1}$$

Then $\hat{f}$ is a continuous function on $\hat{G}$ and trivially satisfies

$$\|\hat{f}\|_{L^\infty(\hat{G})} \le \|f\|_{L^1(G)}. \tag{3.2}$$

The analog of formulae (1.2) and (2.2) asserted that the Haar measure on $\hat{G}$ could be normalized in such a way that

$$\|\hat{f}\|_{L^2(\hat{G})} = \|f\|_{L^2(G)} \tag{3.3}$$

holds for every $f \in L^1(G) \cap L^2(G)$. Then by using the convexity theorem referred to above one can obtain from (3.2) and (3.3) the Hausdorff-Young theorem for locally compact Abelian groups:

$$\|\hat{f}\|_{L^{p'}(G)} \le \|f\|_{L^p(G)}, \quad f \in L^1(G) \cap L^2(G),\ 1 < p < 2 \text{ and } \frac{1}{p} + \frac{1}{p'} = 1. \tag{3.4}$$

In the 1940's and 1950's, the foundations were laid for the study of abstract harmonic analysis on locally compact groups which are neither Abelian nor compact. It was clear that such a study would be useful and that it would have to be contained in the unitary representation theory of locally compact groups on Hilbert spaces of arbitrary infinite dimensions (theorems of Gelfand Raikov (1943) [7] and Freudenthal Weil (1936) [6]). One of the fundamental ideas here was to replace the character group $\hat{G}$ by the collection (also denoted $\hat{G}$) of (equivalence classes of) continuous irreducible unitary representations of G and to consider the family of operator valued integrals $\pi(f) = \int_G f(x)\pi(x)dx$, where π runs over $\hat{G}$, as the Fourier transform of the Haar integrable function f. An essentially equivalent approach based on a consideration of the regular representation of G was to regard the operator L_f of convolution on the left by f on the Hilbert space $L^2(G)$ as the Fourier transform of f. One of the principal tools for developing this approach, aside from the theory of Lie groups and Lie algebras, is the theory of algebras of operators on Hilbert space (C^* - algebras and W^* - algebras (= von Neumann algebras)) which was initiated by Murray and von Neumann in 1929.

In the context of the theory of operator algebras R. Kunze [9] extended the interpolation methods described above so as to prove a Hausdorff-Young Theorem for any unimodular group. This theorem took the form

$$\|L_f\|_{p'} \leq \|f\|_p \text{ for } 1 < p < 2,\ \frac{1}{p} + \frac{1}{p'} = 1. \tag{3.5}$$

Here the norm $\|L_f\|_{p'}$ was defined by using a generalized trace canonically constructed from the group. This result was new even for compact (non-Abelian) groups and reduced to the usual Hausdorff-Young theorem in case G was Abelian. This is so because if G is Abelian, $\hat{f}$ exists and the operator L_f on $L^2(G)$ is unitarily equivalent to the operator $M_{\hat{f}}$ of multiplication by $\hat{f}$ on $L^2(\hat{G})$ and the norm of L_f defined by the trace turns out in this case to be the norm of $\hat{f}$ in the space $L^{p'}(\hat{G})$, as seen in §1 and §2 in the cases $G = \mathbb{T}$ and $G = \mathbb{R}$.

4. Sharpness. In this section we consider the question of whether 1 is the best constant in the inequalities (1.4), (1.6), (2.4) and their generalizations (3.4) and (3.5). First of all it is easy to check that equality holds in (1.4) for any p and for any character on $\mathbb{T}$ i.e. $f(\lambda) = \lambda^n$ for some $n \in \mathbb{Z}$. Therefore 1 is the best constant in (1.4). A similar but more far reaching statement applies to (3.4) and (3.5): there is a non-zero function $f \in L^p(G)$ such that equality holds in (3.5) if and only if G has a compact open subgroup (see [10] and [7 :§43]). This implies that we never have equality in (2.4) for any $p \in (1,2)$ and any $f \neq 0$ in $L^p(\mathbb{R})$. However a much stronger statement holds: there exist constants A_p, $p \in (1,2)$ such that $0 < A_p < 1$ and such that

$$\|\hat{f}\|_{L^{p'}(\mathbb{R})} \leq A_p \|f\|_{L^p(\mathbb{R})},\ \frac{1}{p} + \frac{1}{p'} = 1. \tag{4.1}$$

The smallest such constant is given by

$$A_p = [(p)^{1/p}/(p')^{1/p'}]^{\frac{1}{2}}. \tag{4.2}$$

This was proved first by Babenko [1] in 1961 for the infinite sequence $p = 2m/(2m-1)$ and for arbitrary $p \in (1,2)$ by Beckner [2] in 1975 by different methods. Neither Babenko nor Beckner made use of interpolation methods which would only give the constant 1.

The work of Beckner (for which he was awarded the Salem prize) completely settled the question of the best constant in (3.4) because of the well known structure theorem for Abelian groups. There remains the question of the best constant in (3.5).

The result mentioned above concerning compact open subgroups was improved by Fournier [5] who showed that the best constant in (3.5) is 1 if and only if the unimodular group G has a compact open subgroup. Aside from these two results the question of the best constant in (3.5) for non-Abelian non-compact groups G is being handled on a group by group basis. The interested reader is referred to the author's publications [10,11,12,13] for the details of the results obtained so far, some of which will now be briefly described.

For a locally compact group G let $A_p(G)$ denote the best constant in the Hausdorff-Young theorem for G. Thus $A_p(G) \leq 1$ [9], $A_p(G) = 1$ if and only if G has a compact open subgroup [5], and $A_p(\mathbb{R}) = A_p$[2] where A_p is given by (4.2). The following results are due to the author:

(4.3) $$A_p(\mathbb{R} \times H) = A_p(\mathbb{R})A_p(H) \text{ for all } p \in (1,2) \quad [10]$$

where H is an arbitrary unimodular group.

(4.4) $$A_p(\mathbb{R}^n \times_s H) = A_p(\mathbb{R})^n A_p(H) \text{ for } p' \text{ even} \quad [10]$$

where $A \times_s H$ denotes semidirect product with A normal and H compact

(4.5) $$A_p(\Gamma_3) \leq A_p(\mathbb{R})^2 \text{ for all } p \in (1,2) \quad [11]$$

where Γ_3 is the three dimensional Heisenberg group

(4.6) $$A_p(G_0) \leq A_p(\mathbb{R}) \text{ for all } p \in (1,2) \quad [12]$$

where G_0 denotes the "ax + b" group

(4.7) $$A_p(G) \leq A_p(\mathbb{R}) \text{ for } p' \text{ even} \quad [13]$$

where G denotes any separable Moore group.

In (4.6) a new definition is required since G_0 is not unimodular. In (4.7) use is made of direct integral decompositions and an extension of (1.5) to the case of operator valued kernels [6]. The proof given in §2 is modeled after the proof of (4.7).

Finally in very recent work A. Klein and the author [8] have improved (4.5) and (4.6) to the following

(4.8) $$A_p(\Gamma_3) = A_p(\mathbb{R})^3 \text{ for } p' \text{ even}$$

(4.9) $$A_p(G_0) \leq A_p(\mathbb{R})^2 \text{ for } p' \text{ even.}$$

It seems reasonable at this point to make two conjectures: for all $p \in (1,2)$

(4.10) $A_p(G)$ is always an integral power of $A_p(\mathbb{R})$

(4.11) $$A_p(SL(2,\mathbb{R})) = A_p(\mathbb{R})^2$$

REFERENCES

I. F. Riesz, Uber orthogonale Funktionensysteme, Nachr. Akad. Wiss. Gottingen Math. Phys. Kl. 1907, 116-122.

II. E. Fischer, Sur la convergence en moyenne, C.R. Acad. Sci. Paris 144 (1907) 1022-1024.

III. W. H. Young, On the multiplication of successions of Fourier constants, Proc. Roy. Soc. London, Ser. A 87(1912) 331-339

IV. W. H. Young, on The determination of the summability of a function by means of its Fourier constants, Proc. Lon. Math. Soc. (2) 12(1913) 71-88.

V. F. Hausdorff, Eine Ausdehuung des Parsevalschen Satzes uber Fourier-reihen, Math. Z. 16(1923) 163-169.

VI. M. Riesz, Sur les maxima des formes bilineaires et sur les fonctionnelles lineaires, Acta Math. 49(1926) 465-497.

VII. M. Plancherel, Contribution a letude de la representation d une fonction arbitraire par les integrales definies, Rend. Circ. Mat. Palermo 30(1910) 289-335.

VIII. E. C. Titchmarsh, A contribution to the theory of Fourier transforms, Proc. London Math. Soc. (2) 23(1924) 279-289.

1. K. I. Babenko, An inequality in the theory of Fourier integrals, Izv. Akad. Nauk, SSSR Ser. Mat. 25(1961) 531-542. Engl. Transl. AMS Transl. (2) 44 115-128.
2. W. Beckner, Inequalities in Fourier analysis, Ann. of Math. 102(1975) 159-182.
3. J. Dixmier, "Les algebres d'operateurs dan l'espace Hilbertien", Gauthier Villars, Paris 1957.
4. J. Dixmier, "Les C*-algebres et leurs representations", Gauthier Villars, Paris 1964.
5. J. J. F. Fournier, Sharpness in Young's inequality for convolution, (to appear)
6. J. J. F. Fournier and B. Russo, Abstract interpolation and operator valued kernels, (to appear)
7. E. Hewitt and K. A. Ross, "Abstract Harmonic analysis", Springer-Verlag I (1963), II (1970).
8. A. Klein and B. Russo, Young's inequality for semi-direct products, in preparation.
9. R. A. Kunze, L^p-Fourier transforms on locally compact unimodular groups, Trans. Amer. Math. Soc. 89(1958) 519-540.
10. B. Russo, The norm of the L^p-Fourier transform on unimodular groups, Trans. Amer. Math. Soc. 192(1974) 293-305.
11. _________, The norm of the L^p-Fourier transform. II, Can. J. Math 28(1976)
12. _________, On the Hausdorff-Young Theorem for integral operators, Pacif. J. Math. (to appear)
13. _________, The norm of the L^p-Fourier transform III, (compact extensions), to appear.
14. A. Weil, L'integration dans les groupes topologiques et ses applications, Actualities Sci. et Ind. 869, 1145 Paris: Hermann & Cie. 1941 and 1951.

CLUSTER SETS AND CORONA THEOREMS

Stephen Scheinberg
Department of Mathematics
University of California
Irvine, CA 92264

This note contains the observation that a certain cluster set criterion is equivalent to the corona theorem for algebras of bounded analytic functions, whereas special cases of the criterion are not.

Let Δ, $\overline{\Delta}$, H^∞, and A be respectively the open unit disc, the closed unit disc, the algebra of all bounded analytic on Δ, and the algebra of all continuous functions on $\overline{\Delta}$ which are analytic on Δ. A convenient reference for the basic facts about H^∞ and A is [2]. Regard H^∞ as an algebra of continuous functions on M, the maximal ideal space, and as usual denote by M_λ the set of those maximal ideals (homomorphisms) which contain $z - \lambda$ (send z to λ). If $f = (f_1, \cdots, f_n)$ is an n - tuple of elements of H^∞, denote by R(f) and C(f; λ) the following subsets of $\mathbb{C}^n$: R(f) = the closure of $f(\Delta)$; $C(f;\lambda) = \bigcap_{\epsilon > 0}$ closure of $\{f(z): |z| < 1 \text{ and } |z - \lambda| < \epsilon\}$. An immediate consequence of the corona theorem for H^∞[1] is that R(f) is the range of f on M and that C(f; λ) is the range of f on M_λ.

If B is a subalgebra of H^∞ and n is a positive integer, say that B satisfies I(n) if for every $f_1, \cdots, f_n \in B$ satisfying $\inf_\Delta |f_1| + \cdots + |f_n| > 0$ there exist $g_1, \cdots, g_n \in B$ so that $g_1 f_1 + \cdots + g_n f_n = 1$. Say that B satisfies R(n) if for every $f_1, \cdots, f_n \in B$ and every complex homomorphism φ of B $\varphi(f_1,\cdots,f_n) \in R(f_1,\cdots,f_n)$. Say that B satisfies C(n) if for every such φ there is a $\lambda \in \overline{\Delta}$ such that for every $f_1,\cdots,f_n$ of B $\varphi(f_1,\cdots,f_n) \in C(f_1,\cdots,f_n;\lambda)$. To say that the corona theorem holds for B means, of course, that the image of Δ, under the natural map $\lambda \to$ evaluation at λ, is dense in the maximal ideal space M(B).

<u>Theorem 1</u>. Let B be a uniformly closed subalgebra of H^∞ with $1 \in B$. For each n, I(n) is equivalent to R(n). If $B \supseteq A$ the following are equivalent:

(a) The corona theorem is true for B.

(b) $\forall n$ I(n).

(c) $\forall n$ R(n).

(d) $\forall n$ C(n).

<u>Proof</u>: If R(n) fails, then there are $f_1,\cdots,f_n \in B$ and φ: B $\to$ C with $(\varphi f_1,\cdots,\varphi f_n) \notin R(f_1,\cdots,f_n)$. Thus, $0 \notin R(f_1 - \varphi f_1,\cdots,f_n - \varphi f_n)$. So $\Sigma|f_j - \varphi f_j| \geq \delta > 0$. If I(n) were true we could find $g_1,\cdots,g_n \in B$ so that $\Sigma g_j(f_j - \varphi f_j) = 1$. However, application of φ to this equation yields 0 = 1.

Conversely, if I(n) fails there are $f_1,\cdots,f_n \in B$ with $\Sigma|f_j| \geq \delta > 0$ and no

$\{g_j\}$ exists in B for which $\Sigma g_j f_j = 1$. Therefore, the ideal generated by $f_1, \cdots, f_n$ is proper and so is contained in a maximal ideal. The corresponding homomorphism violates R(n).

Now assume $z \in B$.

(a) $\Rightarrow$ (d). Let φ: $B \to \mathbb{C}$ be a homomorphism and choose a net $z_\alpha \to \varphi$ in the weak - B - topology. Thus, $\lim_\alpha f(z_\alpha) = \varphi(f)$ for each $f \in B$. Since $z \in B$, we can define $\lambda = \varphi(z)$; $|\lambda| \leq 1$ and $\lim z_\alpha = \varphi(z) = \lambda$. Clearly, $\lim_\alpha (f_1(z_\alpha), \cdots, f_n(z_\alpha)) = (\lim_\alpha f_1(z_\alpha), \cdots, \lim_\alpha f_n(z_\alpha)) = (\varphi f_1, \cdots, \varphi f_n)$ is an element of $C((f_1, \cdots, f_n); \lambda)$. It is obvious that (d) $\Rightarrow$ (c).

(c) $\Rightarrow$ (a): If Δ is not dense in M(B), choose $\varphi \in M(B)$ which is not in $\widetilde{\Delta}$, the closure of Δ in M(B). Because of the nature of the topology of M(B), we can find (for some n) $f_1, \cdots, f_n \in \ker \varphi$ such that $\Sigma |f_j| \geq \delta > 0$ on Δ. This contradicts R(n).

<u>Theorem 2</u>. For each n there exists a uniformly closed subalgebra $B_n \subseteq H^\infty$, $B_n \supseteq A$, such that B_n satisfies C(n) (and hence C(k), R(k), and I(k) for every $k \leq n$) but not the corona theorem.

<u>Remark</u>: To my knowledge the algebras B_n are the first known examples of uniform algebras between A and H^∞ for which the corona theorem fails.

<u>Proof</u>: The idea of the proof is to obtain, given n, a function algebra A_n with the property that each n - tuple of its elements has the same range on a proper closed $X \subseteq M(A_n)$ as on all of $M(A_n)$. A suitable subset Y of M_1 ($\subseteq M(H^\infty)$) is mapped onto X and the algebra B_n is defined as the set of those elements of H^∞ which restricted to Y correspond via the mapping to elements of A_n. Homomorphisms of B_n are determined in a natural way by homomorphisms of H^∞ or by homomorphisms of A_n. In either case an n - tuple of elements of B_n is mapped into its range on M, which by the corona theorem for H^∞ is its cluster set. However, because Y is proper in $M(A_n)$, Δ is not dense in $M(B_n)$. We now turn to a more precise rendering of these ideas.

Let G be an open ball or polydisc in $\mathbb{C}^{n+1}$ and let X be the topological boundary of G. Let A_n be the algebra of continuous functions on $G \cup X$ which are analytic on G. $M(A_n) = G \cup X$ and every function achieves its maximum on X. It is well known that if $\mathcal{O}$ is an open set in $\mathbb{C}^m$ and V is an analytic variety in $\mathcal{O}$ of dimension at least 1, then $\overline{V}$ meets $\partial\mathcal{O}$. From this it follows that every n - tuple of elements of A_n has the same range on X as on $G \cup X$.

Enumerate a dense subset $x_1, x_2, \cdots$ of X and fix a bijection between the natural numbers N and N × N. Let t be the first coordinate of this mapping and define $\dot{\alpha}$: $N \to X$ by $\dot{\alpha}(n) = x_{t(n)}$. α extends to a unique continuous map, which we again call α, of βN onto X. In fact, α maps $N^* = \beta N - N$ onto X.

Identify $n \in N$ with $z_n = 1 - 2^{-n} \in \Delta$. Since $\{z_n\}$ is an interpolating sequence for H^∞, we obtain an embedding of β N into M, via which N^* corresponds to a closed subset Y of M_1. So we obtain a map γ of Y onto X. $\gamma^* f = f \circ \gamma$ defines an isometry of $C(X)$ into $C(Y)$. The restriction map ρ of H^{∞} into $C(Y)$ is norm decreasing; it is onto because z_n is interpolating. Define B_n to be $\rho^{-1}\gamma^*(A_n)$. $(\gamma^*)^{-1}\rho$ is a norm-decreasing map of B_n onto A_n; so completeness of A_n implies completeness of B_n. The kernel of $(\gamma^*)^{-1}\rho$ is clearly I, the set of elements of B_n which vanish identically on Y; $B_n/I \cong A_n$. Next we establish that if $\varphi: B_n \to \mathbb{C}$ is a homomorphism, either φ extends to be a homomorphism on H^{∞} or else φ factors through ρ, as in the following diagram:

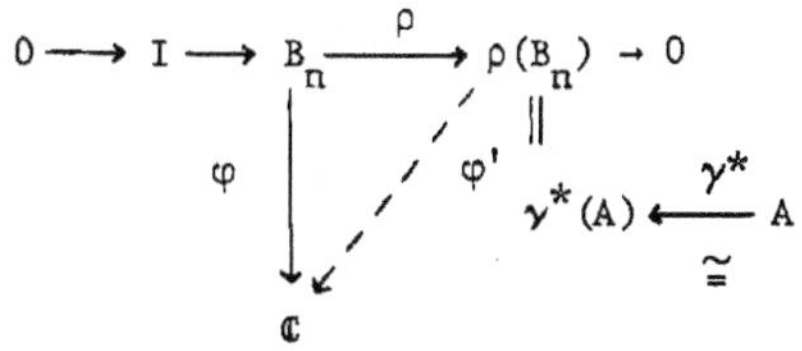

Indeed, if $\varphi^{-1}(0) \supseteq I$, then φ factors through ρ because the (top) row of the diagram is exact. Otherwise, there is an element $i \in I$ such that $\varphi(i) \neq 0$. We may assume $\varphi(i) = 1$ and then define $\tilde{\varphi}: H^{\infty} \to \mathbb{C}$ by $\tilde{\varphi}(f) = \varphi(fi)$. This makes sense because $H^{\infty}i \subseteq I$. $\tilde{\varphi}$ is linear and $\tilde{\varphi}(i) = 1$. $\tilde{\varphi}(fg) = \varphi(fgi) = \varphi(i)\ \varphi(fgi) = \varphi(if)\ \varphi(gi) = \tilde{\varphi}(f)\tilde{\varphi}(g)$. So $\tilde{\varphi}$ is the desired extension to H^{∞}.

Now it is easy to see that $\varphi(f_1,\cdots,f_n) \in C(f_1,\cdots,f_n; \varphi z)$. For if φ extends to $\tilde{\varphi}$, we know $\tilde{\varphi} = \lim z_{\alpha}$ and this assertion is obvious. And if φ factors through ρ, $\varphi = \varphi'\rho$, then $\varphi'\gamma^*$ is a homomorphism of A_n. That is, for each $f \in B_n$ there is a $g \in A_n$ so that $\varphi f = \varphi'\rho f = \varphi'\gamma^* g$. Given $f_1,\cdots,f_n$ we find $(\varphi f_1,\cdots,\varphi f_n) = (\varphi'\gamma^* g_1,\cdots,\varphi'\gamma^* g_n)$ is an element of the range of $(g_1,\cdots,g_n)$ on $G\cup X$, which is the range of $(g_1,\cdots,g_n)$ on X, which is contained in the range of $(\gamma^* g_1,\cdots,\gamma^* g_n)$ on Y, which is contained in the range of $(f_1,\cdots,f_n)$ on M_1, as desired.

Finally, to see that the corona theorem is violated for B_n, let $f_1,\cdots,f_{n+1}$ be elements of B which correspond (via $(\gamma^*)^{-1}\rho$) to the coordinate functions $Z_1,\cdots,Z_{n+1}$ on $\mathbb{C}^{n+1}$, where we assume $0 \in G$, as we may. Inf $\sum_1^{n+1} |Z_j| > 0$ on X means $\inf \sum_1^{n+1} |f_j| > 0$ on $Y \cup \{z_k: k \geq K\}$, for some large K. Put $f_0 = \prod_K^{\infty}\left(\frac{z_k - z}{1 - \bar{z}_k z}\right)$. Because $\{z_k\}$ is an interpolating sequence, the zeros of f_0 in M are exactly (see [3]) the closure of $\{z_k: k \geq K\}$, which is $Y \cup \{z_k: k \geq K\}$. Therefore, $f_0 \in B_n$ and $\sum_0^{n+1} |f_j| > 0$ on M; by compactness of M $\inf \sum_0^{n+1} |f_j| > 0$ on Δ. However, the homomorphism of B_n corresponding to $0 \in G$ annihilates $f_0,\cdots,f_{n+1}$.

REFERENCES

1. L. Carleson, Interpolation by bounded analytic functions and the corona problem, Ann. of Math. 76(1962), pp. 542-559.

2. K. Hoffman, Banach Spaces of Analytic Functions, Prentice Hall, Englewood Cliffs, N.J., 1962.
3. K. Hoffman, Bounded analytic functions and Gleason parts, Ann. of Math. 86(1967), pp. 74-111.

REMARKS ON F-SPACES OF ANALYTIC FUNCTIONS

Joel H. Shapiro*
Mathematics Department
Michigan State University
East Lansing, MI 48824

and

Mathematics Department
University of Wisconsin-Madison
Madison, WI 53706

1. Introduction. An F-space is a complete, metrizable linear topological space. This article is intended as a case study showing how F-spaces which are not locally convex generate interesting questions which do not occur in the locally convex theory. When these questions are related to concrete spaces of analytic functions they lead to new problems in function theory, and conversely the spaces of analytic functions provide examples that are interesting for the general theory.

The non locally convex phenomenon of interest here is the existence of proper, closed subspaces that are dense in the weak topology. According to the Hahn-Banach theorem such objects never occur in the locally convex theory. In section 2 of this paper we will see that they can occur in F-spaces which are not locally convex: whether they always occur is an open problem. Sections 3 and 4 discuss how these subspaces show up among the shift-invariant subspaces of the Hardy spaces H^p for $0 < p < 1$, and how they can be used to construct interesting examples of F-spaces with trivial dual. The fifth section contains a more detailed examination of some of these matters. In particular there is a fairly complete proof that $H^{\frac{1}{2}}$ contains a proper, closed, weakly dense invariant subspace, and there is a discussion of the "Banach envelope" of $H^p(0 < p < 1)$. The final section deals with the Hardy Algebra N^+, a classical space of analytic functions closely related to the Hardy spaces, whose weakly dense invariant subspaces can be completely characterized.

2. Separation properties. We are going to focus on the way in which continuous linear functionals on an F-space separate points from each other and from closed subspaces. Suppose E is a linear topological space with (topological) dual E'. We say E has the:

a) Point separation property if E' separates points of E, or equivalently if for every $0 \neq x$ in E there is a non-trivial continuous linear functional λ on E with $\lambda(x) \neq 0$;

---

*I want to thank the Department of Mathematics, University of Wisconsin-Madison for its hospitality during the preparation of this paper.

b) Hahn-Banach separation property if E' separates points of E from closed subspaces not containing them, or equivalently if every closed subspace of E is weakly closed;

c) Hahn-Banach approximation property if each proper, closed subspace of E is annihilated by some non-trivial continuous linear functional, or equivalently if every weakly dense subspace is dense.

It is easy to see that the last two properties can be characterized in terms of quotient spaces; E has the Hahn-Banach separation property iff every quotient of E (by a closed subspace) has the point separation property; and the Hahn-Banach approximation property iff no nontrivial quotient has trivial dual.

Clearly the Hahn-Banach separation property implies the other two. For locally convex (Hausdorff) spaces the Hahn-Banach theorem guarantees all of them, but a space that is not locally convex may have none of them. Perhaps the most famous example of such pathology is the F-space $L^p = L^p([0,1])$ for $0 < p < 1$, taken in its natural metric $d(f,g) = \|f - g\|$, where

$$\|f\| = \int_0^1 |f(t)|^p dt \qquad (f \text{ in } L^p). \tag{2.1}$$

In 1940 M.M. Day [2] showed that L^p has no non-trivial continuous linear functionals, i.e. $(L^p)' = \{0\}$ for $0 < p < 1$. In particular, L^p has none of the above separation properties.

Before we continue, note that the functional $\|\cdot\|$ defined by (2.1) is subadditive on L^p, but not homogeneous. Instead it is p-homogeneous:

$$\|af\| = |a|^p \|f\|$$

for each f in L^p and each scalar a. Such functionals (subadditive, p-homogeneous, vanishing only at the origin) are called p-norms.

An example more subtle than L^p is the sequence space ℓ^p $(0 < p < 1)$ with the natural topology induced by the p-norm

$$\|f\| = \sum_1^\infty |f(n)|^p \qquad (f = (f(n))_1^\infty \text{ in } \ell^p).$$

For each positive integer n the evaluation functional

$$\lambda_n(f) = f(n) \qquad (f \text{ in } \ell^p)$$

is continuous, and the family $(\lambda_n)_1^\infty$ separates points of ℓ^p, so ℓ^p has the point separation property. However, ℓ^p does not have the Hahn-Banach approximation property (so it does not have the Hahn-Banach separation property either) when $0 < p < 1$ [27], [30]. This is not difficult to see. First recall the result of Mazur and Orlicz which states that every separable Banach space is isomorphic to a quotient of ℓ^1. Essentially the same proof (let (e_n) be the standard unit vector basis of ℓ^1, choose a countable dense subset (x_n) of the unit ball of the Banach space, extend the bijection $e_n \to x_n$ by linearity and continuity to all of ℓ^1, prove that the extended map is onto) shows that for $0 < p < 1$ every separable p-normed F-space is isomorphic to a quotient of ℓ^p. In particular L^p, which has trivial dual, is isomorphic to a quotient of ℓ^p, so ℓ^p fails to have the Hahn-Banach approximation property by the quotient space characterization.

The closed subspace-call it K- by which ℓ^p was divided to get L^p is an example of a proper, closed, weakly dense subspace in ℓ^p. Unfortunately K is not an easy object to lay hands on: all we get from the proof of its existence is that

$$K = \{f \text{ in } \ell^p : \sum_1^\infty f(n)x_n = 0\}$$

for (x_n) a fixed dense subset of the L^p unit ball (e.g. (x_n) = all trigonometric polynomials with rational coefficients). So other than the fact that K exists, it does not seem to lead to any interesting analysis in ℓ^p.

The situation is different for the Hardy spaces H^p. In 1969 Duren, Romberg, and Shields studied these spaces for $0 < p < 1$ and found proper, closed, weakly dense subspaces <u>invariant under multiplication by the independent variable</u> z. Unlike the subspace K of the previous paragraph, these invariant subspaces of H^p are tractable from the point of view of function theory, and the fact that some of them are weakly dense when $0 < p < 1$ raises new questions about the structure of H^p functions. This matter will be the subject of the next section.

Motivated by these examples, Duren, Romberg, and Shields posed two general questions:

(1) Does every non locally convex F-space fail to have the Hahn-Banach separation property?

(2) Does every non locally convex F-space fail to have the Hahn-Banach approximation property?

The first question was recently answered in the affirmative by N. J. Kalton:

<u>KALTON'S THEOREM</u> [12; Cor. 5.3]. <u>An F-space is locally convex if and only if it has the Hahn-Banach separation property</u>.

An equivalent statement in the language of quotient spaces is this: <u>If an F-space is not locally convex</u>, <u>then some quotient has a dual which does not separate points</u>.

The second question is still open. An equivalent formulation is: <u>does every non locally convex F-space have a nontrivial quotient with trivial dual</u>?

Finally, it should be pointed out that replacing "F-space" by "Hausdorff linear topological space" in these two questions changes the situation drastically. In fact there are many non locally convex linear topological spaces with the Hahn-Banach separation property, and hence also the Hahn-Banach approximation property. For example any real or complex vector space of uncountable Hamel dimension endowed with its strongest vector topology has this property (see [15; p. 53, Problem 6I] and [5; pp. 59-60]). A different class of examples arises from the fact if E is an infinite dimensional Banach space, then E supports a non locally convex topology $\mathcal{T}$ intermediate between its weak and norm topologies [7]. Clearly the non locally convex space (E, $\mathcal{T}$) has the Hahn-Banach separation property (by the Hahn-Banach theorem).

3. Weakly dense invariant subspaces in H^p $(0 < p < 1)$. This section introduces the "nice" examples of proper, closed, weakly dense subspaces that occur in the Hardy spaces, and indicates some of the function theoretic problems they suggest. We begin with a brief review of some basic theory: a good reference for this material is the first three chapters of Duren's book [3].

Definition of H^p. The Hardy space H^p is the collection of functions f analytic in the open unit disc Δ such that

$$\|f\|_p^p = \sup_{0 \leq r < 1} \frac{1}{2\pi} \int_{-\pi}^{\pi} |f(re^{it})|^p \, dt < \infty.$$

The case usually studied is $1 \leq p < \infty$, where $\|\cdot\|_p$ is a norm which makes H^p into a Banach space. However the interest here is in the range $0 < p < 1$, where the functional $\|\cdot\|_p^p$ is a p-norm which makes H^p into an F-space [3; Page 37, Cor. 2]. That H^p fails to be locally convex was first noticed by Livingston [17]*. For each point z in Δ, the evaluation functional

$$\lambda_z(f) = f(z) \qquad\qquad (f \text{ in } H^p)$$

is continuous [3; Chapter 7, Page 118], and the family $(\lambda_z : z \text{ in } \Delta)$ separates the points of H^p. Thus H^p has the point separation property, even when $0 < p < 1$.

The boundary correspondence. An important link between H^p theory and real analysis is provided by a theorem of Fatou which states that for each f in H^p the radial limit

$$f^*(e^{it}) = \lim f(re^{it}) \qquad (r \to 1-)$$

exists for almost every t, and the boundary function f^* belongs to $L^p(T)$ (T = unit circle) with

$$\frac{1}{2\pi} \int_{-\pi}^{\pi} |f^*(e^{it})|^p \, dt = \|f\|_p^p \tag{3.1}$$

[3; Theorem 2.6, P. 21]. In other words the "boundary correspondence" $f \to f^*$ is a

*Regarding this, note that any Banach space can masquerade as a p-normed space: if $\|\cdot\|$ is the norm, then $\|\cdot\|^p$ is a p-norm inducing the same topology.

linear isometry taking H^p onto a closed subspace of $L^p(T)$. The boundary function f^* even retains a vestige of analyticity in that it cannot be too small too often. More precisely, if $0 \neq f \in H^p$, then [3; Theorem 2.2, Page 17]

$$\int_{-\pi}^{\pi} \log|f^*(e^{it})|\,dt > -\infty. \tag{3.2}$$

This implies, for example, that f^* cannot vanish on a subset of T having positive measure.

<u>Inner-outer factorization</u>. Along with some classical function theory, the boundary correspondence yields an important structure theorem for H^p functions. We can ease into this result by noting that for f fixed in H^p, (3.1) and (3.2) imply that $\log|f^*|$ is integrable on T. Thus the formula

$$F(z) = \exp\{\frac{1}{2\pi}\int_{-\pi}^{\pi} \frac{e^{it}+z}{e^{it}-z} \log|f^*(e^{it})|\,dt\} \tag{3.3}$$

defines a function F analytic on Δ and without zeroes there. Moreover F belongs to H^p [3; Section 2.4]. Now $\log|F(z)|$ is the Poisson integral of $\log|f^*|$, so standard facts about Poisson integrals and subharmonic functions [3; Chapter 1] show that $\log|f| \leq \log|F|$ on Δ and $\log|f^*| = \log|F^*|$ almost everywhere on T. Thus $q = f/F$ is a function analytic on Δ (since F never vanishes), bounded in modulus by 1, with $|q^*(e^{it})| = 1$ a.e.. Such a function q is called an <u>inner function</u>, and F as given by (3.3) is called an <u>outer function</u>. Thus we see that <u>every</u> H^p <u>function</u> f <u>can be factored as</u>

$$f = qF \tag{3.4}$$

<u>where</u> q <u>is inner and</u> F <u>is outer</u>. It turns out that this factorization is unique up to a multiplicative constant of modulus 1.

A rough interpretation of this inner-outer factorization might go as follows: the outer factor carries the modulus of the boundary function (since $|f^*| = |F^*|$ a.e.), and the inner factor carries the zeroes of the "interior" function (since F never vanishes). But this is not quite the whole story: the inner factor q itself has a further decomposition which is probably best understood through some functional analysis.

<u>Invariant subspaces</u>. Let U denote the operator of "multiplication by z" on H^p, that is:

$$(Uf)(z) = zf(z) \qquad (f \text{ in } H^p,\ z \text{ in } \Delta).$$

U is often called the <u>right shift</u>, or <u>unilateral shift</u> because it shifts the Taylor coefficients of f one unit to the right. U is noteworthy because it is one of the few infinite dimensional bounded operators whose invariant subspace structure has been completely worked out. This was done for p = 2 (i.e. Hilbert space) by Beurling, whose theorem forms the basis for a lot of modern interest in H^p spaces.

BEURLING'S THEOREM [1], [11; Ch. 7, p. 99], [10; Lecture II]. A closed subspace of H^p is invariant under U (henceforth just "invariant") if and only if it has the form qH^p for some inner function q.

Clearly each subspace qH^p is invariant, and since $|q^*| = 1$ a.e. it follows from the boundary correspondence that qH^p is closed. So the force of Beurling's theorem lies in the other direction: every invariant subspace is "generated" by an inner function. The cases $p \neq 2$ follow from the case $p = 2$: for $1 \leq p < \infty$ the arguments are given in Helson's monograph [10; Lecture IV], and the case $0 < p < 1$ goes just like $1 \leq p < 2$.

Beurling's theorem can also be viewed as a result on approximation. In this formulation it states that the polynomial multiples of an H^p function form a dense subset of H^p if and only if that function is outer [3; Section 7.3]. In 1969 Duren, Romberg, and Shields added a new dimension to Beurling's result by proving that when $0 < p < 1$ some inner functions (not identically 1) give rise to weakly dense invariant subspaces in H^p. In view of the approximation-theoretic formulation of Beurling's theorem, Duren, Romberg, and Shields called such inner functions weakly outer. It is a challenging unsolved problem in function theory to characterize in a useful way the inner functions that are weakly outer. In particular it is not known if an inner function can be weakly outer for some values of $0 < p < 1$, but not for others.

In order to get some feeling for how the "weakly outer" phenomenon can happen, we need to look at some examples of inner functions and the invariant subspaces they generate.

Examples of inner functions and invariant subspaces.

(a) Blaschke factors. These are functions of the form

$$B(z) = \frac{\alpha - z}{1 - \bar{\alpha} z}$$

for α a fixed complex number in Δ. So B is just the conformal mapping of Δ onto itself which interchanges 0 and α. Clearly B is an inner function, and it is easy to see that the invariant subspace BH^p is precisely the collection of H^p functions that vanish at α. More generally, if n is a positive integer, then B^n is an inner function and B^nH^p is just the collection of H^p functions with a zero of order $\geq n$ at α.

(b) Blaschke products. If (z_n) is a sequence of points in Δ which satisfies the Blaschke condition $\Sigma(1 - |z_n|) < \infty$, then the Blaschke product

$$B(z) = \Pi \frac{z_n}{|z_n|} \frac{z_n - z}{1 - \bar{z}_n z}$$

certainly converges at the origin, and in fact it converges uniformly on each compact subset of Δ [3; Theorem 2.4]. Since each factor in the product is an inner function (a Blaschke factor rotated to have positive value at the origin) it seems reasonable to hope that B itself will be inner, and this is exactly what happens. The corresponding invariant subspace BH^p then consists entirely of H^p functions vanishing on the sequence (z_n), with multiplicity at each point $\geq$ the number of times that point occurs in the sequence ($< \infty$ by the Blaschke condition). Moreover, a factorization theorem we will state shortly shows that BH^p consists of all such functions. It follows that for B a Blaschke product, the invariant subspace BH^p is weakly closed in H^p, even when $0 < p < 1$. This is most easily seen when each member of the sequence (z_n) occurs exactly once, for then $BH^p = \bigcap_n \lambda_{z_n}^{-1}(0)$, where the evaluation functionals λ_{z_n} are continuous. For the general case, evaluation of f at z_n is replaced by evaluation of an appropriate derivative of f at z_n (still a continuous linear functional). So no Blaschke product can be weakly outer.

(c) Singular inner functions. There are also inner functions with no zeroes in Δ. Perhaps the most notorious of these is the unit singular function

$$S(z) = \exp\left\{\frac{z+1}{z-1}\right\}$$

which is clearly inner because $\frac{z+1}{z-1}$ is a conformal map taking Δ onto the left half-plane. While the unit singular function has no zeroes in Δ, it has in a certain sense a very strong "boundary zero," in that it decays very quickly to 0 along the unit interval. More precisely:

(3.5) $$S(r) = O(e^{-1/(1-r)})$$

as $r \to 1-$. Now the growth of each f in H^p is limited by the condition

(3.6) $$|f(z)| = O((1-r)^{-1/p}) \qquad (r \to 1-,\ z = re^{it})$$

[3; Theorem 5.9] so each function in the invariant subspace SH^p decays to zero along the unit interval at essentially the same exponential rate as S, hence has the same kind of "boundary zero." We might conjecture from this that SH^p is weakly closed in H^p when $0 < p < 1$; and this is precisely the case (although the proof, which will be indicated in Section 5, proceeds along different lines).

The unit singular function can be generalized in the following way. Let μ be a positive measure on T singular with respect to Lebesgue measure. Then the function

$$S_\mu(z) = \exp\left\{\int \frac{z+e^{it}}{z-e^{it}}\, d\mu(t)\right\}$$

is an inner function (since $\log|S(re^{it})| = -$Poisson integral of μ, hence is ≤ 0

on Δ, and $\to 0$ a.e. as $r \to 1-$) which has no zeroes in Δ. S_μ is called the singular inner function generated by μ. For example the unit singular function is S_δ, where $d\delta(t)$ = unit mass at $t = 0$.

Singular inner functions are more subtle objects than Blaschke products in that they generate invariant subspaces not associated with zeroes of functions and derivatives. On the other hand every inner function can be factored into a Blaschke product and a singular inner function [3; Theorem 2.8], the factorization being unique up to multiplication by a complex number of modulus one. This result, and the previous factorization (3.4) show that every H^p function can be factored into an inner factor which carries the zeroes (namely the Blaschke product), an outer factor which carries the boundary modulus, and a singular inner factor which has properties in common with both the outer factor (it never vanishes in Δ), and the Blaschke product (it is inner). Of course in order to make these statements correct, we must allow the function $\equiv 1$ to be simultaneously the trivial Blaschke product, singular inner function, and outer function.

The weakly dense invariant subspaces ($0 < p < 1$). Here at last is the result we are aiming for. The preceding discussion shows that if an inner function is going to be weakly outer (i.e. generate a weakly dense invariant subspace) in $H^p(0<p<1)$, then it cannot have zeroes So by the factorization theorem just stated, it must be a singular inner function. But the unit singular function, which generates a weakly closed invariant subspace, shows that more is needed. Intuitively, the measure generating the unit singular function is so rough that it induces radial decay in the invariant subspace that persists even after weak closure. So the question is: can a measure on T be singular, yet smooth enough that the corresponding singular inner function generates a weakly dense invariant subspace? Duren, Romberg, and Shields [5] proved that such measures exist. In order to state their result properly we need a way of determining the smoothness of a measure.

Definition. The modulus of continuity of a finite positive Borel measure on T is the function

$$\omega_\mu(\delta) = \sup_{|I| \leq \delta} \mu(I) \qquad (\delta > 0)$$

where I runs through the intervals of T, and $|I|$ is the length of I.

Note that $\omega_\mu(\delta) = 0(\delta)$ implies that μ is absolutely continuous with respect to Lebesgue measure. However it is well known that there exist positive singular measures with any preassigned modulus of continuity rougher than $0(\delta)$; for example $\omega_\mu(\delta) = 0(\delta \log \frac{1}{\delta})$ occurs as such a modulus of continuity [24],[4],[9].

The result of Duren, Romberg, and Shields is the following:

THEOREM [5; Theorem 13, Page 53]. Suppose μ is a positive singular measure on T, with modulus of continuity $\omega_\mu(\delta) = O(\delta \log \frac{1}{\delta})$. Then the invariant subspace $S_\mu H^p$ is weakly dense in H^p, that is, S_μ is weakly outer for $0 < p < 1$.

The proof of this result will be sketched in the fifth section: for the case $p = \frac{1}{2}$ it will be reasonably complete. For the moment it is enough for us to know that H^p contains proper, closed, weakly dense subspaces that are naturally connected with the analysis of the space, when $0 < p < 1$. We have already seen that this leads to an interesting problem in function theory: characterize those singular measures μ whose associated inner functions S_μ are weakly outer. Now we are going to show that the existence of these "nice" weakly dense subspaces leads to interesting examples in the theory of F-spaces.

4. Some F-spaces with trivial dual. I want to show how the results of the previous section can be used to solve a problem in the general theory of F-spaces. The problem, due to A. Pelczynski, is this: can an F-space with trivial dual have non-trivial compact endomorphisms? Another way of asking this is: can an F-space with no nontrivial finite rank endomorphisms have a nontrivial compact endomorphism? Recently Pallaschke [18] showed that L^p, which has trivial dual, actually has no non-trivial compact endomorphism ($0 < p < 1$), and Kalton [13] generalized this by showing that there is no non-trivial compact operator from L^p into any Hausdorff linear topological space. However, using the existence of proper, closed, weakly dense invariant subspaces in H^p ($0 < p < 1$), Kalton and I were able to show that there exist F-spaces with trivial dual which nevertheless have non-trivial compact endomorphisms [14]. Here is the idea of the proof.

Fix $0 < p < 1$, and let κ denote the topology of uniform convergence on compact subsets of Δ (restricted to H^p). The hero of this section is going to be τ, the strongest topology on H^p that agrees on bounded subsets of H^p with κ. That is, we declare a set to be τ-open if and only if its intersection with every bounded set B is relatively κ-open in B. It is easy to check that these τ-open sets really are the open sets for a topology, and it is not difficult to see that τ is Hausdorff and weaker than the original topology of H^p. What is not so easy is to check that τ is a vector topology, but fortunately it is. Two other properties of τ which are not difficult to prove are:

(1) The closed unit ball of H^p is τ-compact,

and

(2) the invariant subspace qH^p is τ-closed for every inner function q. It is the last property that is most critical: we would not be able to prove it if qH^p were not such an explicitly described subspace.

Now choose q to be a singular inner function which generates a weakly dense (proper) invariant subspace qH^p. The existence of such inner functions was discussed in the last section. Then the quotient $E = H^p/qH^p$ is an F-space with trivial dual. On the other hand, since qH^p is $\mathcal{T}$-closed, the quotient $F = H^p/qH^p$ in the quotient $\mathcal{T}$-topology is a Hausdorff linear topological space with a topology weaker than that of E (since $\mathcal{T}$ is weaker on H^p than the original topology). In particular, F has trivial dual, and the identity map $E \to F$ is continuous. In fact it is even compact, since the identity map $H^p \to (H^p, \mathcal{T})$ is compact (due to the $\mathcal{T}$-compactness of the H^p unit ball). Thus the map $E + F \to E + F$ defined by $(e,f) \to (0,e)$ is a non-trivial compact endomorphism of $E + F$, and $E + F$ is a linear topological space which has trivial dual, since E and F have trivial dual.

So the fact that the invariant subspace qH^p is weakly dense, but $\mathcal{T}$-closed, leads rather directly to a linear topological space $E + F$ with trivial dual, which nonetheless supports non-trivial compact endomorphisms. Unfortunately $E + F$ is not metrizable (since it turns out that F isn't), but it is not difficult to modify the construction and replace F by a metrizable space (see [14] for the details), yielding an F-space with trivial dual but non-trivial compact endomorphisms. q.e.d.

This gives one instance of how H^p theory can furnish examples that are interesting in the study of F-spaces. Another such example is the quotient space $E = H^p/qH^p$; the first factor in the direct sum mentioned above. We have just seen that E is an F-space with trivial dual, that there is a compact operator taking E into a Hausdorff linear topological space (namely the identity map $E \to F$), and that there is no such operator on L^p. So E is not isomorphic to L^p: in fact E is a new F-space with trivial dual, and should be interesting to study in its own right. For example, does E itself have a compact endomorphism? Does E have a compact convex subset with no extreme point? James Roberts [19] has recently shown that there are linear topological spaces with such subsets: in fact he has shown that L^p is such a space for $0 < p < 1$ [20]. At the moment it is not clear if every linear topological space with trivial dual must contain such a set.

5. The Banach envelope and weakly outer inner functions. This section explores the "weakly outer" phenomenon in greater detail. Surprisingly the controlling interest in the problem belongs to a Banach space of analytic functions that contains H^p.

The Banach envelope. Suppose E is a p-normed F-space ($0 < p \leq 1$) with the point separation property. Then a simple exercise shows that the convex hull of the unit ball of E contains no linear subspaces, so its Minkowski functional [15; Page 15] is a norm on E which induces a topology weaker than the original one (equal to it iff E was locally convex to begin with) but having the same continuous linear functionals. The completion of E in this norm is a Banach space $\hat{E}$ called the Banach envelope of E. Thus E is contained in $\hat{E}$, the inclusion map is continuous, and the dual of E is the same as that of $\hat{E}$ in the sense that the restriction map $\lambda \to \lambda|_E$ takes $\hat{E}'$ onto E'.

It is instructive to follow through this construction for the case $E = \ell^p$ ($0 < p < 1$). Since $\|f\|_1 \leq \|f\|_p$ for all f in ℓ^p we see that ℓ^p is continuously embedded in ℓ^1 via the identity map, so ℓ^1 is a candidate for $\hat{\ell}^p$. In fact $\hat{\ell}^p = \ell^1$ in the sense that the identity map $\ell^p \to \ell^1$ extends to an isometry of $\hat{\ell}^p$ onto ℓ^1. To see this we need only show that the ℓ^1-closure of the convex hull of the ℓ^p-unit ball is the ℓ^1-unit ball. This is easy. Let (e_n) denote the standard unit vector basis for ℓ^1. Then for each f in ℓ^1 the partial sums of the series representation $f = \Sigma f(n)e_n$ are convex combinations of elements in the ℓ^p unit ball, and they converge in ℓ^1 to f.

The fact that E and its Banach envelope have the same dual shows that a subspace of E is weakly dense if and only if it is dense in $\hat{E}$. So when dealing with specific examples, one way to show that a subspace is weakly dense is to get hold of a concrete representation of the Banach envelope and show that the subspace is dense therein. For H^p $(0 < p < 1)$ such a representation was obtained by Duren, Romberg, and Shields:

THEOREM [5; Section 3]. The Banach envelope of H^p is the space B_p of functions f analytic in Δ such that

$$\|f\|_{B_p} = \iint_\Delta |f(z)|(1 - |z|)^{1/p-2}\, dxdy < \infty .$$

Of course the theorem is stated rather loosely: it really means that H^p is contained in B_p, the identity map $H^p \to B_p$ is continuous, and it extends to an isomorphism (not isometry this time) of $\hat{H}^p$ onto B_p. To be precise we should perhaps say that ℓ^1 is an isometric representation of the Banach envelope of ℓ^p, while B_p is merely an isomorphic representation of the Banach envelope of H^p $(0 < p < 1)$. In any case the important thing for our purposes is that a singular inner function q is weakly outer if and only if qH^p (or equivalently the polynomial multiples of q) is dense in B_p. So the weakly outer phenomenon in H^p is also an approximation problem in a certain Banach space of analytic functions. Such problems have been studied extensively by H. S. Shapiro [23], [25], [26], and his results play an important role in our story. Before getting on with this, however, I would like to indicate why B_p is an isomorphic representation of $\hat{H}^p$, at least for $p = \frac{1}{2}$.

$\hat{H}^{\frac{1}{2}} = B_{\frac{1}{2}}$. Note that $B_{\frac{1}{2}}$ is just the subspace of $L^1(\Delta)$ consisting of analytic functions. Our first task is to show that $H^{\frac{1}{2}}$ is contained in $B_{\frac{1}{2}}$ and the identity map is continuous, i.e. that $\|f\|_{B_{\frac{1}{2}}} \le C\|f\|_{\frac{1}{2}}$ for some constant C independent of $f \in H^{\frac{1}{2}}$. This is a result of Hardy and Littlewood [8; Theorem 31]: a short proof can be based on inner-outer factorization and Hardy's inequality.

$$\Sigma(n + 1)^{-1}|\hat{f}(n)| \le \pi\|f\|_1 \tag{5.1}$$

for each $f(z) = \Sigma\hat{f}(n)z^n$ in H^1 [3; Page 48].

Since $|\hat{f}(n)| \leq \|f\|_1$ for f in H^1, Hardy's inequality yields

$$\frac{1}{\pi}\iint_{\Delta}|f(z)|^2\,dxdy = \Sigma(n+1)^{-1}|\hat{f}(n)|^2 \leq \pi\|f\|_1^2 \tag{5.2}$$

for f in H^1. Now if $f \in H^{\frac{1}{2}}$ and <u>has no zeroes</u> in Δ, then $f^{\frac{1}{2}} \in H^1$ and $\|f^{\frac{1}{2}}\|_1 = \|f\|_{\frac{1}{2}}^{\frac{1}{2}}$. So replacing f by $f^{\frac{1}{2}}$ in (5.2) we obtain

$$\|f\|_{B_{\frac{1}{2}}} = \iint_{\Delta}|f(z)|\,dxdy \leq \pi^2\|f\|_{\frac{1}{2}}, \tag{5.3}$$

which is the desired inequality. In case f has zeroes in Δ we can still apply this result by using a trick due to Hardy and Littlewood. Let $f = qF$ be the inner-outer factorization (3.4) of f, let $G = (q-1)F$, so $f = G + F$ where neither G nor F have zeroes in Δ, and $\|G\|_{\frac{1}{2}} \leq 2\|f\|_{\frac{1}{2}}$; while as we observed in Section 3, $\|F\|_{\frac{1}{2}} = \|f\|_{\frac{1}{2}}$. Applying (5.3) to F and G respectively yields:

$$\begin{aligned}\|f\|_{B_{\frac{1}{2}}} &\leq \|G\|_{B_{\frac{1}{2}}} + \|F\|_{B_{\frac{1}{2}}}\\ &\leq \pi^2(\|G\|_{\frac{1}{2}} + \|F\|_{\frac{1}{2}}) \qquad \text{(by (5.3))}\\ &\leq 3\pi^2\|f\|_{\frac{1}{2}},\end{aligned}$$

which completes the proof that $H^{\frac{1}{2}}$ is contained in $B_{\frac{1}{2}}$ with the identity map continuous.

To complete the proof that $\hat{H}^{\frac{1}{2}}$ is isomorphically represented by $B_{\frac{1}{2}}$ we need to show that the closure in $B_{\frac{1}{2}}$ of the $H^{\frac{1}{2}}$ unit ball contains some ball in $B_{\frac{1}{2}}$. The idea is similar to the one used for ℓ^p, except that instead of using a basis to represent elements of $B_{\frac{1}{2}}$ we use the <u>reproducing kernel</u>

$$K(z,\zeta) = (\beta + 1)\,\frac{(1-|\zeta|^2)^{\beta}}{(1-\bar{\zeta}z)^{\beta+2}}\,.$$

A calculation with power series shows that if $\beta > 0$ and $f \in B_{\frac{1}{2}}$ then

$$f(z) = \iint_{\Delta}K(z,\zeta)f(\zeta)dA(\zeta) \qquad (z \text{ in } \Delta) \tag{5.4}$$

where $dA(\zeta)$ is normalized Lebesgue area measure on Δ. So if we write $K(\zeta)(z) = K(z,\zeta)$ then each $K(\zeta)$ is an analytic function in z on the closed unit disc, and we can think of (5.4) as representing each f in the $B_{\frac{1}{2}}$ unit ball as a sort of generalized absolutely convex combination of these functions. In fact the right side of (5.4) really is a limit in $B_{\frac{1}{2}}$ of absolutely convex combinations of $K(\zeta)$'s, namely the approximating Riemann sums for the integral (see [28; Section 3] for the details), while a straightforward calculation with $\beta = 2$ shows that $\|K(\zeta)\|_{\frac{1}{2}} \leq 3$ for all ζ in Δ. This shows that the $B_{\frac{1}{2}}$ unit ball lies in the $B_{\frac{1}{2}}$-closure of the convex hull of the $H^{\frac{1}{2}}$ ball of radius 3 about the origin, and completes the proof that the Banach envelope of $H^{\frac{1}{2}}$ is (isomorphically represented by) $B_{\frac{1}{2}}$. For other values of p the details of the last half of the proof are similar, with a few more exponents to keep straight [28].

Existence of weakly outer inner functions. We can now outline a proof of the Duren, Romberg, Shields theorem which states that a singular inner function S_μ is weakly outer whenever its modulus of continuity is $O(\delta \log \frac{1}{\delta})$. For simplicity we give the proof only for $p = \frac{1}{2}$, although it is hardly more difficult for the other values of $0 < p < 1$. The argument is due to H. S. Shapiro [26; Theorem 1].

The essential point is that the modulus of continuity condition on the singular measure μ is equivalent to the growth condition

$$m(r) = \min_{|z|=r} |S_\mu(r)| \geq C(1-r)^N \tag{5.5}$$

for some positive constants C and N independent of $0 \leq r < 1$ (see [23, Theorem 2] for the details). Thus S_μ decays to zero rather slowly near the boundary; and given our remarks about the unit singular function we appear to be on the right track. In fact if $N < 1$ we are finished, since taking p_n to be the nth arithmetic mean of the Taylor series of $1/S_\mu$ we have $p_n \to 1/S$ uniformly on compact subsets of Δ, and [16; Kap. 1, Satz 1, p. 22]

$$\max_{|z|=r} |p_n(z)| \leq \max_{|z|=r} |1/S_\mu(z)| \leq 1/m(|z|) \leq C^{-1}(1-|z|)^{-N}$$

Thus

$$|S_\mu(z)p_n(z)| \leq C^{-1}(1-|z|)^{-N},$$

where the right side is integrable over Δ, and $S_\mu p_n \to 1$ pointwise on Δ. So the Lebesgue Dominated Convergence theorem implies that $S_\mu p_n \to 1$ in $B_{\frac{1}{2}}$. Adopting the notation [E] for the closure in $B_{\frac{1}{2}}$ of the subset E, we see from the above that $[S H^{\frac{1}{2}}]$ contains all polynomials. But the polynomials are dense in $B_{\frac{1}{2}}$ [3, Theorem 3] so $[S_\mu H^{\frac{1}{2}}] = B_{\frac{1}{2}}$, and the proof is complete for $N < 1$.

In case $N \geq 1$ note that $S_\mu^{1/2N}$ is the singular inner function induced by the measure $\mu/2N$, and it obeys estimate (5.5) with exponent $= \frac{1}{2}$. So the last paragraph shows that $[S_\mu^{1/2N} H^{\frac{1}{2}}] = B_{\frac{1}{2}}$. Now it is easy to see that if q_1 and q_2 are inner functions with q_2 weakly outer, then $[q_1 q_2 H^{\frac{1}{2}}] = [q_1 H^{\frac{1}{2}}]$, so taking $q_1 = q_2 = S_\mu^{1/2N}$ we obtain

$$\begin{aligned} [S_\mu^{1/N} H^{\frac{1}{2}}] &= [S_\mu^{1/2N} S_\mu^{1/2N} H^{\frac{1}{2}}] \\ &= [S_\mu^{1/2N} H^{\frac{1}{2}}] \\ &= B_{\frac{1}{2}}. \end{aligned}$$

Repeating the argument N times yields $[S_\mu H^{\frac{1}{2}}] = B_{\frac{1}{2}}$. QED

<u>The unit singular function is not weakly outer</u>. We close this section by sketching the proof that the unit singular function $S(z) = \exp \frac{z+1}{z-1}$ generates an invariant subspace that is not weakly dense.

So we have to find a non-trivial continuous linear functional on H^p that vanishes on SH^p. A device for writing down continuous linear functionals on H^p is furnished by the estimate

$$(5.6) \qquad |\hat{f}(n)| \leq C\, n^{1/p-1} \|f\|_p \quad ,$$

where C is independent of the function $f(z) = \Sigma \hat{f}(n) z^n \in H^p$, and $0 < p < 1$ [3; Theorem 6.4, Page 98]. For example if $(a_n)_0^\infty$ is a sequence of complex numbers with

$$(5.7) \qquad |a_n| = 0(n^{-1-1/p})$$

then it follows from (5.6) that the formula

$$(5.8) \qquad \lambda(f) = \sum_0^\infty a_n \hat{f}(n) \qquad (f \text{ in } H^p)$$

defines a continuous linear functional on H^p. We will be done if we can find a non-trivial sequence (a_n) satisfying (5.7) and such that the linear functional λ given by (5.8) annihilates SH^p. The way to do this is to study the function

$$h(e^{it}) = e^{it}(1 - e^{it})^k\, \overline{S(e^{it})} \sim \sum_{-\infty}^{\infty} \hat{h}(n) e^{int} ,$$

which can be made to have as many continuous derivatives as desired simply by choosing k sufficiently large ($k = 2n + 1$ guarantees n continuous derivatives, to be precise). Choosing k so that h has at least $1 + 1/p$ continuous derivatives we obtain the estimate $|\hat{h}(n)| = 0((|n| + 1)^{-1-1/p})$. Let $a_n = \hat{h}(-n)$ for $n \geq 0$, so (a_n) obeys (5.7), and λ as defined by (5.8) is continuous on H^p. We claim that λ annihilates SH^p. To see this observe that λ can also be expressed by the integral formula

$$\lambda(f) = \frac{1}{2\pi} \int_{-\pi}^{\pi} f^*(e^{it}) h(e^{it}) dt$$

at least for f a polynomial. Recalling that for a.e. t we have $|S^*(e^{it})| = 1$, the integral formula for λ gives for each polynomial f:

$$\lambda(Sf) = \frac{1}{2\pi} \int_{-\pi}^{\pi} f(e^{it})(1-e^{it})^k\, e^{it}\, dt = 0 .$$

Since the polynomials are dense in H^p [3; Theorem 3.3], the polynomial multiples of S are dense in SH^p, so λ vanishes on all of SH^p. So the proof is complete, provided $a_n = \hat{h}(-n) \neq 0$ for some $n > 0$. If this were not the case then h would be the boundary function of an H^2 function g. Now $e^{it}(1-e^{it})^k$ is the boundary function of the H^2 function $f(z) = z(1-z)^k$, and

$$(Sg)^* = S^*g^* = S^*h = f^* \text{ a.e.},$$

hence the H^2 function SG - f has radial limits zero a.e.. But in Section 3 we remarked (after inequality (3.2)) that a non-trivial H^2 function cannot have 0 as a radial limit on a set of positive measure. Thus Sg = f. But this is impossible: for example (3.5) and (3.6) would then guarantee that $f(r) = r(1-r)^k$ tends to zero exponentially fast as $r \to 1-$, which is clearly absurd. Thus $a_n \neq 0$ for some positive n, and the proof is complete.

This argument is due to H. S. Shapiro [25; Theorem 2], who actually proved that S is not weakly outer whenever μ gives positive measure to some thin set. Here a closed subset K of the unit circle is called thin if $\Sigma\, \alpha_n \log(1/\alpha_n) < \infty$, where α_n is the length of the nth interval in $T \setminus K$. The case we have just treated is the one where K is a singleton.

6. Weakly dense closed ideals in the Hardy Algebra. In this section we completely characterize the weakly outer inner functions in the Hardy Algebra, a space of analytic functions which is in some sense the "limit" of the H^p spaces as $p \to 0+$.

Let N denote the space of functions f analytic in Δ and having bounded characteristic:

$$\|f\| = \sup_{0 \le r < 1} \frac{1}{2\pi} \int_{-\pi}^{\pi} \log(1+|f(re^{it})|dt < \infty, \tag{6.1}$$

and let N^+ denote those functions f in N for which the family $\{\log^+|f(re^{it})| : 0 \le r < 1\}$ is uniformly integrable on the unit circle T. N is called the Nevanlinna Class, while N^+ is the Hardy Algebra [6; Ch. 5, Sec. 2] or Smirnov Class [3; Sec. 2.5 and Page 31]. Both N and N^+ are algebras under pointwise multiplication, and $N \supset N^+ \supset H^p$ for all $p > 0$. The metric induced by the subadditive functional $\|\cdot\|$ makes N into a complete space in which N^+ is a closed subspace. Surprisingly this metric does not make N into a linear topological space - the scalar multiplication is discontinuous [29], but fortunately N^+ is spared this embarassment. It turns out that N^+ is a complete topological algebra in the metric induced by $\|\cdot\|$; and that while N^+ is not locally convex, it does have the point separation property (for example it is not difficult to show that the point evaluations $f \to f(z)$ are continuous for each z in Δ).

The Hardy Algebra is often regarded as the limiting case of the H^p spaces because it shares many properties with H^p; in particular the boundary correspondence, the inner-outer factorization, Beurling's theorem, and the density of polynomials (cf., [31; Theorem 4]). Beurling's theorem becomes particularly attractive in this setting. Since N^+ is a topological algebra in which polynomials are dense, a closed subspace is invariant under multiplication by z if and only if it is an ideal. Thus Beurling's theorem for N^+ states that the closed ideals are precisely those of the form qN^+ where q is an inner function, that is, the closed ideals are the principal ideals generated by inner functions.

Since N^+ is not locally convex it is possible that some closed ideals are weakly dense. Using the methods of Section 5 these can be characterized as follows:

THEOREM. A closed ideal in N^+ is weakly dense if and only if it is generated by a singular inner function (i.e. an inner function without zeroes.)

This result can be restated in a couple of ways: an ideal is weakly dense iff it has no common zero in Δ; or an inner function is weakly outer in N^+ iff it is singular. In any case the continuity of point evaluations shows that a closed weakly dense ideal can only be generated by a singular inner function, so the task at hand is to show that every singular inner function does generate such an ideal (recall that the analogous result for H^p was false). This has been done by Roberts and Stoll [21] for the case where the measure associated with the inner function has no point mass.

To prove the theorem we need an analogue for N^+ of the Banach envelope. Since the function $\|\cdot\|$ is not p-homogeneous for any $0 < p \le 1$ this requires some care, but the idea is still the same. Let C_n denote the convex hull of the N^+ - ball of radius 1/n about the origin. It is easy to check that the family (C_n) of convex sets forms a local base for a locally convex metrizable topology on N^+ that is weaker than the original one, but has the same continuous linear functionals [15; Page 109]. The completion $\hat{N}^+$ of N^+ in this topology is a Fréchet space (locally convex F-space) containing N^+ and having the same dual: we might call it the Fréchet envelope of N^+. Just as in the previous section, a subspace of N^+ is weakly dense if and only if it is dense in the Fréchet envelope.

So a concrete representation of $\hat{N}^+$ is needed. This was recently supplied by N. Yanagihara [31] who identified $\hat{N}^+$ with the space F^+ of functions f analytic in Δ such that

$$(6.2) \qquad \sup |f(z)| \exp\{-c/(1-|z|\} < \infty \qquad (|z| < 1)$$

for every $c > 0$. The precise statement is that $F^+ \supset N^+$, and the identity map $N^+ \to F^+$ extends to an isomorphism of F^+ onto N^+, where F^+ has the topology induced by the seminorms (6.2) as c runs through all positive reals. It is not difficult to see that the seminorms (6.2) can be replaced by an equivalent family

$$(6.3) \qquad \|f\|_c = \iint_\Delta |f(z)| \exp\{-c(1-|z|)\}\, dxdy$$

where c runs through the positive reals (in fact Yanagihara actually defined F^+ in terms of still a third equivalent family of seminorms).

Proof of the Theorem (Cf. [26; Theorem 1]). Let q be a singular inner function. We must show that the ideal qN^+ is dense in F^+, or equivalently that the polynomial multiples of q are dense in the normed space $(F^+, \|\cdot\|_c)$ for each $c > 0$. Since

$|q(z)| = \exp\{-\text{Poisson integral of } \mu\}$, where μ is the singular measure that induces q, and since the Poisson integral of μ at z is $\leq 2\|\mu\|/(1-|z|)$, it follows that the minimum modulus of q is limited by

$$m(r) = \min_{|z|=r} |q(z)| \geq \exp\{-2\|\mu\|/(1-r)\}.$$

Now proceeding as in the last section, let p_n be the nth arithmetic mean of the partial sums of the Taylor series of $1/q$, so $p_n q \to 1$ pointwise on Δ, and for each z in Δ:

$$|q(z)\, p_n(z)| \leq |p_n(z)| \leq 1/m(|z|) \leq \exp\{2\|\mu\|/(1-|z|)\}.$$

so when $c \geq 2\|\mu\|$ the Dominated Convergence Theorem shows that $\|p_n q - 1\|_c \to 0$, hence qN^+ is dense in $(F^+, \|\cdot\|_c)$. For $c < 2\|\mu\|$ the above argument still works on q^ϵ for $0 < \epsilon \leq c/2\|\mu\|$, and an iteration like the one performed in the last section shows that qN^+ is dense in $(F^+, \|\cdot\|_c)$, which completes the proof.

REFERENCES

1. A. Beurling, On two problems concerning linear transformations in Hilbert space. Acta Math. 81(1949), 239-255.
2. M. M. Day, The spaces L^p with $0 < p < 1$. Bull. Amer. Math. Soc. 46(1940), 816-823.
3. P. L. Duren, Theory of H^p Spaces. Academic Press, New York, 1970.
4. ________, Smoothness of functions generated by Riesz products. Proc. Amer. Math. Soc. 16(1965), 1263-1268.
5. P. L. Duren, B. W. Romberg, and A. L. Shileds, Linear functionals on H^p spaces with $0 < p < 1$. J. reine angew. Math. 238(1969), 32-60.
6. T. W. Gamelin, Uniform Algebras. Prentice Hall, Englewood Cliffs, N.J. 1969.
7. D. A. Gregory and J. H. Shapiro, Nonconvex linear topologies with the Hahn-Banach extension property. Proc. Amer. Math. Soc. 25(1970), 902-905.
8. G. H. Hardy and J. E. Littlewood, Some properties of fractional integrals II. Math. Z. 34(1932), 403-439.
9. P. Hartman and R. Kershner, The structure of monotone functions. Amer. J. Math. 59(1937), 809-822.
10. H. Helson, Lectures on Invariant Subspaces. Academic Press, New York, 1964.
11. K. Hoffman, Banach Spaces of Analytic Functions. Prentice-Hall, Englewood Dliffs, N.J. 1962.
12. N. J. Kalton, Basic sequences in F-sapces and their applications. Proc. Edinburgh Math. Soc. (2)19(1974), 151-167.
13. ________, Compact and strictly singular operators on Orlicz spaces. Israel J. Math. (to appear).
14. N. J. Kalton and J. H. Shapiro, An F-space with trivial dual and non-trivial compact endomorphisms. Israel J. Math. 20(1975), 282-291.
15. J. L. Kelley, I. Namioka, et al., Linear Topological Spaces, Van Nostrand, Princeton, 1963.
16. E. Landau, Darstellung und Begrundung einiger neurer Ergebnisse der Funktiontheorie. Springer, Berlin, 1929.
17. A. E. Livingston, The space H^p, $0 < p < 1$, is not normable. Pacific J. Math. 3(1953), 613-616.
18. D. Pallaschkte, The compact endomorphisms of the metric linear space ϕ. Studia Math. 47(1973), 153-165.
19. J. W. Roberts, A compact convex set with no extreme points. Studia Math., to appear.
20. ________, In preparation.
21. J. W. Roberts and M. Stoll, Prime and principal ideals in the algebra N^+. Arkiv der Matematik, 27(1976), 387-393.

22. W. Rudin, Real and Complex Analysis, McGraw-Hill, New York, 1966.
23. H. S. Shapiro, Weakly invertible elements in certain function spaces, and generators in ℓ^1. Michigan Math. J. 11(1964), 161-165.
24. _________, A class of singular functions. Canad. J. Math. 20(1968), 1425-1431.
25. _________, Some remarks on weighted polynomial approximation of holomorphic functions. Mat. Sbornik 73(1967), 320-330 (in Russian).
26. _________, Weighted polynomial approximation and boundary behavior of holomorphic functions. In the volume "Contemporary Investigations in the Theory of Functions", Nauka, Moscow, 1966, 326-335.
27. J. H. Shapiro, Examples of proper, closed, weakly dense subspaces in some F-spaces of analytic functions. Israel J. Math. 7(1969), 369-380.
28. _________, Mackey topologies, reproducing kernels, and diagonal maps on the Hardy and Bergman spaces. Duke Math. J. 43(1976), 187-202.
29. J. H. Shapiro and A. L. Shields, Unusual topological properties of the Nevanlinna Class. Amer. J. Math. 97(1976), 915-936.
30. W. J. Stiles, On properties of subspaces of ℓ_p, $0 < p < 1$. Trans. Amer. Math. Soc. 149(1970), 405-415.
31. N. Yanigahara, The containing Fréchet space for the class N^+. Duke Math. J. 40(1973), 93-103.
32. _________, Multipliers and linear functionals for the class N^+. Trans. Amer. Math. Soc. 180(1973), 449-461.

EXTREMELY SMOOTH BANACH SPACES

Mark A. Smith
Department of Mathematics
Lake Forest College
Lake Forest, IL 60045/USA

Francis Sullivan
Department of Mathematics
The Catholic University of America
Washington, D.C. 20064/USA

1. Introduction and background. Let X be a Banach space and X*, X**, X*** and $X^{(4)}$ the successive dual spaces. In 1948 Dixmier [7] proved that if X is not reflexive then $X^{(4)}$ is not strictly convex. Later, Rainwater [11], using James' characterization of weak compactness, showed that if X*** is smooth then X is reflexive. In [15] it was observed that Rainwater's proof consists of showing that the points of S* (the unit sphere of X*) which do not attain their norm on S (the unit sphere of X) are not smooth points of S***. In other words if X* is "very smooth" then X is reflexive. Giles proved that norm to weak continuity of the norming map from S to S* characterizes very smooth" then X is reflexive. Giles proved that norm to weak continuity of the norming map from S to S* characterizes very smooth spaces [8] and Tacon showed that a dual space of a very smooth space has an equivalent strictly convex norm [16]. It is also known that any space with a Fréchet differentiable norm is very smooth [15] and that a dual space of a very smooth space has the Radon-Nikodym property [6][15]. At the end of this paper we obtain a strengthening of this last result.

The notion of an "extremely smooth" norm was also introduced in [15]. This is a norm for which the norming functionals for a fixed point in S** all agree on X. If X** is smooth then clearly X is both very smooth and extremely smooth. In [13] it was shown that if $\dim(X^{**}/X) \leq 1$ then the converse also holds, namely X** is smooth. This gives an example of a non-reflexive space having smooth second dual. Moreover, in this case if the norm in X is uniformly Gateaux differentiable and X* is locally uniformly convex then X*** is strictly convex [14]. This gives an example of a strictly convex space with a nonsmooth dual that is separable. In this paper we obtain an improvement of the latter result in terms of properties related to "extremely smooth".

In [15] it was proved that if the norm in X is uniformly Gateaux differentiable then X is extremely smooth, and that for an extremely smooth space the norming map from S to S* can be extended in a natural way to a map from S** to B* (the unit ball of X*). The main purpose of the present paper is to investigate more fully extremely smooth norms and other geometrical properties related to uniform Gateaux differentiability of the norm.

Day showed that ℓ^∞ has no equivalent smooth norm [3]. His proof is based, in part, on the properties of the set of norm-1 vectors in the metric complement of c_0, i.e. $\{x \in \ell^\infty : \|x\| = 1 \text{ and } \mathrm{dist}(x,c_0) = 1\}$. Our approach to obtaining information on the degree of smoothness in the general situation is also based on properties of the metric complement of X in X**. We show that if X is extremely smooth, then for any $x^{**} \in X^{**}$, $\mathrm{dist}(x^{**},X) = \|x^{**}\|$ if and only if every norming sequence for x** converges weak* to zero. Using the same ideas we obtain some quantitative information on how far a non-reflexive space must diverge from being uniformly smooth.

We show that "extremely smooth" is implied by a property expressed in terms of increasing nested families of balls. Notions of this sort provide a different viewpoint on smoothness and have been used by Vlasov [19] in connection with approximation theoretic questions. A similar notion related to "very smooth" was considered in [15].

We introduce some other new geometrical properties and establish their relationships to "uniformly Gateaux differentiable" and "extremely smooth". As consequences of the characterization of the metric complement of X in X** for extremely smooth X we obtain criteria expressed in terms of "extremely smooth" and these related notions that imply reflexivity.

Our main results are then summarized in a chart depicting the implications that exist among the various properties. This form of presentation seems to be the most convenient way of indicating the relative strengths of these notions.

We conclude the paper with the improvements of the results in [6], [14] and [15] mentioned above.

The remainder of this section consists of definitions and other background material.

A Banach space X is said to be <u>smooth</u> if for each $x \in S$ there exists a unique $x^* \in S^*$ such that $x^*(x) = 1$. If (x_k^*) and (y_k^*) are norm-1 sequences and X is smooth, then if for any $x \in S$, $x_k^*(x)$, $y_k^*(x) \to 1$ we have $x_k^* - y_k^* \overset{*}{\to} 0$. A stronger condition is that X be <u>extremely smooth</u>, namely; if (x_k^*) and (y_k^*) are norm-1 sequences such that $x^{**}(x_k^*)$, $x^{**}(y_k^*) \to 1$ for some $x^{**} \in S^{**}$, then $x_k^* - y_k^* \overset{*}{\to} 0$. An equivalent formulation of extremely smooth in terms of Gateaux differentiability is that for all $x^{**} \in S^{**}$ and $y \in S$

$$\lim_{t \to 0^+} \frac{\|x^{**} + ty\| + \|x^{**} - ty\| - 2}{t} = 0.$$

This was proved in [15].

We denote by J_0, J_1 and J_2 the natural embeddings of X, X* and X** into X**, X*** and $X^{(4)}$ respectively. When no confusion can result we shall write simply $x \in X^{**}$ to mean $J_0 x \in X^{**}$. Goldstine's Theorem says that the weak* closure of X in X** is all of X**. Using Helly's Theorem, Lindenstrauss and Rosenthal [9] obtained the following strengthening of this result.

Proposition. Let $A \subseteq X^{**}$ and $F \subseteq X^*$ be finite dimensional subspaces and let $0 < \delta < 1$ be arbitrary. Then there exists a linear map $T:A \to X$ such that:

(a) $T(a) = a$ for all $a \in A \cap X$,

(b) $f(T(a)) = a(f)$ for all $a \in A$ and $f \in F$,

(c) $(1 - \delta)\|a\| \leq \|T(a)\| \leq (1 + \delta)\|a\|$ for all $a \in A$.

This has come to be known as the principle of local reflexivity. The statement given here is due to Dean [5].

It is well known that $X^{***} = X^* \oplus X^\perp$ and X^* is the range of the contractive projection $J_1J_0^*$ on X^{***}. Passing to dual spaces we have that $X^{(4)} = X^{**} \oplus X^{*\perp}$ and X^{**} is the range of $J_2J_1^*$. In $X^{(4)}$ the closed subspace $X^{*\perp}$ is also the null manifold of the contractive projection $(J_1J_0^*)^* = J_0^{**}J_1^*$. It is easy to check that for any $x^{**} \in X^{**}$ we have $J_2x^{**} - J_0^{**}x^{**} \in X^{*\perp}$. Dixmier proved that $\|J_2x^{**} - J_0^{**}x^{**}\| \geq \text{dist}(x^{**},X)$.

2. Extremely smooth norms and related notions. For $x^* \in X^*$ a norm-1 sequence (x_k) in X is called a norming sequence for x^* if $x^*(x_k) \to \|x^*\|$.

Theorem 1. If X is extremely smooth, then for any x^{**} in X^{**}, $\text{dist}(x^{**},X) = \|x^{**}\|$ if and only if every norming sequence for x^{**} converges weak* to zero.

Proof. Assume that $\text{dist}(x^{**},X) = \|x^{**}\| = 1$ and that (x_k^*) is a norming sequence. From weak* compactness and the definition of extremely smooth given in section 1 we have that (x_k^*) converges weak* to, say, x^*. If $\|x^*\| \geq a > 0$ then for some $x \in S$ and all k large enough, $x_k^*(x) \geq a/2 > 0$. We shall show that X can not be extremely smooth at this x.

The fact that $\text{dist}(x^{**},X) = 1$ is equivalent to $x^\perp(x^{**}) = 1$ for some norm-1 $x^\perp \in X^\perp \subseteq X^{***}$. Hence, for any $t > 0$ and all k

$$\frac{\|x^{**} + tx\| + \|x^{**} - tx\| - 2}{t}$$

$$\geq \frac{x^{**}(x_k^*) + tx_k^*(x) + x^\perp(x^{**}) - 2}{t}$$

$$= x_k^*(x) + \frac{x^{**}(x_k^*) - 1}{t}.$$

Thus, if (t_j) is a sequence decreasing to zero, by passing to a subsequence we may assume that $\dfrac{x^{**}(x_j^*) - 1}{t_j} \to 0$ and obtain a contradiction of the fact that X is extremely smooth.

For the converse, if $x^{**} \in S^{**}$ and $x^{**}(x_k^*) \to 1$ where (x_k^*) is a norm-1 sequence converging weak* to zero, then in X^{***} some subset of (x_k^*) converges weak* to, say, x^{***}. Clearly $\|x^{***}\| \leq 1$ and $x^{***} \in X^\perp$. Hence $\text{dist}(x^{**},X) = 1$. QED

Corollary. If X is extremely smooth and dist(x**,X) = 1 for some x** in S**, then X** is not very smooth at x**.

Proof. The Theorem shows that x** can not attain its norm on S* and from the original argument of Rainwater [11], x** is not a very smooth point. QED

Corollary. If X is extremely smooth then the set of x** in S** such that dist(x**,X) < 1 is norm dense in S**.

Proof. From the Theorem dist(x**,X) < 1 for any x** ∈ S** which attains its norm on S*. By the Bishop-Phelps Theorem [1] this set is norm dense in S**. QED

The ideas employed in the previous results can also be used to give a criterion for reflexivity in terms of smoothness properties. First we give some definitions. For $x \in S$ and $t > 0$ let

$$\rho(x,t) \equiv \sup\{\|y + tx\| + \|y - tx\| - 2 : \|y\| = 1\}.$$

Notice that $0 \le \rho(x,t) \le 2t$ for each x and X uniformly Gateaux differentiable at x is just the statement that $\lim_{t\to 0^+} \frac{\rho(x,t)}{t} = 0$. For $t > 0$ define $\rho(t) = \sup\{\rho(x,t) : \|x\| = 1\}$. Then X uniformly smooth is equivalent to $\lim_{t\to 0^+} \frac{\rho(t)}{t} = 0$.

The following theorem generalizes the known fact that if X is uniformly smooth then X is reflexive.

Theorem 2. If $\liminf_{t \to 0^+} \frac{\rho(t)}{t} < 1$ then X is reflexive.

Proof. For $0 \le \epsilon < 1$ and $x \in S$ let $\tau(\epsilon,x) = \sup\{0 < t < 1 : \rho(x,t) \le \epsilon t\}$. We shall show that if X is not reflexive, then for each $0 \le \epsilon < 1$ we can produce a sequence of norm-1 vectors (x_k) such that $\tau(\epsilon,x_k) \to 0$. Hence, if (ϵ_k) is a positive sequence increasing to one, we can choose a sequence (x_k) such that $\bar{t}_k \equiv \tau(\epsilon_k,x_k) \to 0$. Notice that for each k, if $t > \bar{t}_k$ then $\frac{\rho(x_k,t)}{t} > \epsilon_k$. Therefore, if (t_j) is any sequence decreasing to zero, we can choose $\bar{t}_{k_j} < t_j$ for each j, and so

$$\frac{\rho(t_j)}{t_j} \ge \frac{\rho(x_{k_j},t_j)}{t_j} > \epsilon_{k_j}$$ which gives the result.

By the previous reduction we need only show that $\inf\{\tau(\epsilon,x) : \|x\| = 1\} = 0$ for all $0 \le \epsilon < 1$. If for some $0 \le \epsilon < 1$ and all $x \in S$ we have $\tau(\epsilon,x) \ge b > 0$ then $(1 - \epsilon)\tau(\epsilon,x) \ge (1 - \epsilon)b \equiv a > 0$.

Using the principle of local reflexivity as in [15] we have that for any $0 < t < \tau(\epsilon,x)$ and all $\|x^{**}\| = 1$

$$\epsilon \ge \frac{\|x^{**} + tx\| + \|x^{**} - tx\| - 2}{t}.$$

Let $0 < \delta < a/2$. Since X is not reflexive we may choose $\|x^{**}\| = 1$ such that $x^{\perp}(x^{**}) \geq 1 - \delta$ for some $\|x^{\perp}\| = 1$. Moreover, from the Bishop-Phelps Theorem we may assume $x^{**}(x^*) = 1$ for some $\|x^*\| = 1$. Thus

$$\begin{aligned} \epsilon &\geq \frac{x^{**}(x^*) + tx^*(x) + x^{\perp}(x^{**}) - 2}{t} \\ &\geq \frac{1 + tx^*(x) + 1 - \delta - 2}{t} \\ &= x^*(x) - \delta/t. \end{aligned}$$

Since $\tau(\epsilon,x)$ is independent of the choice of x^{**}, x^* and $x^{\perp}$ we have that $\epsilon + \delta/\tau(\epsilon,x) \geq x^*(x)$ for all $\|x\| = 1$, and so

$$\epsilon + \delta \sup\{1/\tau(\epsilon,x) : \|x\| = 1\} \geq 1.$$

It follows that

$$\delta \geq (1 - \epsilon) \inf\{\tau(\epsilon,x) : \|x\| = 1\} \geq a > 2\delta,$$

a contradiction. QED

We now turn our attention to other geometrical properties that are related to "extremely smooth".

A Banach space X is said to be rotund (or strictly convex) if $\|x + y\| = 2$ implies that $x = y$ for norm-1 vectors x and y. It is clear that if X* is rotund then X is smooth and it is known that the implication can not be reversed [17]. In fact rotundity of X* is equivalent to the smoothness of every two dimensional quotient space of X [4; p. 145]. An equivalent formulation of this was given by Vlasov [19].

Theorem 3. (Vlasov) X* is rotund if and only if for every nested sequence

$$B_1 \subseteq B_2 \subseteq \cdots \subseteq B_n \subseteq B_{n+1} \subseteq \cdots$$

of closed balls in X with radii increasing and unbounded, the norm closure of $\bigcup_n B_n$ is either all of X or a half space.

This result has been useful in obtaining information on the convexity of Chebyshev sets [19].

In [15] a stronger smoothness property which is also expressed in terms of an increasing sequence of balls was considered. This was called property V and it was shown to be equivalent to rotundity conditions on X* that are sufficient to imply that X is very smooth. We consider a similar notion expressed in terms of a sequence of finite nested sequences of balls. We denote the closed ball with center at x and radius r by $B(x;r)$.

We say that a Banach space X has property W if there do not exist norm-1 sequences (x_k^*), (y_k^*) and closed balls B_m^n for $n = 1,2,\cdots$ and $1 \leq m \leq n$ such that

(a) $B_m^n \subseteq B_{m+1}^n$ for all $m < n$,

(b) if (j) is an increasing sequence of integers and $n(j)$ is another sequence such that $j \leq n(j)$ for all j, then the radii $(r_j^{n(j)})$ become unbounded,

(c) for some real b, $x_k^*(x) \geq b$ and $y_k^*(x) \geq b$ for all $x \in B_m^n$ where $m \leq k \leq n$,

(d) for some $z \in S$, $(x_k^* - y_k^*)(z) \geq a > 0$ for all k.

Theorem 4. If X has property W then X is extremely smooth.

Proof. Assume that X is not extremely smooth. From the results of [15] there exist $x^* + x^\perp$, $y^* + y^\perp \in S^{***}$ and $x^{**} \in S^{**}$ such that $(x^* + x^\perp)(x^{**}) = (y^* + y^\perp)(x^{**}) = 1$ and $x^* \neq y^*$.

Clearly for some $z \in S$ we have that

$$[(x^* + x^\perp) - (y^* + y^\perp)](z) = (x^* - y^*)(z) \geq a > 0.$$

We shall use the principle of local reflexivity (or really just Helly's Theorem) repeatedly to produce closed balls and functionals contradicting the definition of property W.

First, from local reflexivity we obtain sequences of norm-1 vectors (x_k^*) and (y_k^*) such that for all k

$$\begin{aligned} x^{**}(x_k^*) &\geq 1 - \epsilon_k \\ x^{**}(y_k^*) &\geq 1 - \epsilon_k \end{aligned} \tag{1}$$

where $(k + 1)\epsilon_k < 1$ and $(x_k^* - y_k^*)(z) \geq a/2$. Now, we apply local reflexivity to the inequalities (1) to obtain a norm-1 vector x_1^1 such that

$$x_1^*(x_1^1) \geq 1 - \epsilon_1 - \epsilon_1^1$$

$$y_1^*(x_1^1) \geq 1 - \epsilon_1 - \epsilon_1^1$$

where $\epsilon_1^1 \leq \epsilon_1$. Repeating this we next produce a pair of norm-1 vectors (x_1^2, x_2^2) such that

$$x_1^*(x_1^2), \quad y_1^*(x_1^2) \geq 1 - \epsilon_1 - \epsilon_1^2$$

$$x_2^*(x_1^2), \quad y_2^*(x_1^2) \geq 1 - \epsilon_2 - \epsilon_1^2$$

$$x_2^*(x_2^2), \quad y_2^*(x_2^2) \geq 1 - \epsilon_2 - \epsilon_2^2$$

where $\epsilon_1^2 + \epsilon_2^2 \leq \epsilon_2$.

At step n we obtain norm-1 vectors $(x_1^n, x_2^n, \cdots, x_n^n)$ satisfying the inequalities

$$x_k^*(x_m^n), \qquad y_k^*(x_m^n) \geq 1 - \epsilon_k - \epsilon_m^n$$

for all $m \leq k \leq n$ where $\sum_{m=1}^{n} \epsilon_m^n \leq \epsilon_n$.

Let $B_m^k = B(x_1^k + \cdots + x_m^k; m)$. Properties (a) and (b) for this collection of balls are clear. If $\|u\| \leq m \leq k \leq n$ then

$$\begin{aligned} & x_k^*(x_1^n + \cdots + x_m^n - u) \\ & \quad \geq (1 - \epsilon_k - \epsilon_1^n) + \cdots + (1 - \epsilon_k - \epsilon_m^n) - m \\ & \quad \geq -m\epsilon_k - \sum_{i=1}^{m} \epsilon_i^n \\ & \quad \geq -(m + 1)\epsilon_k > -1. \end{aligned}$$

The same inequality holds for y_k^* and so the proof is complete. QED

We next consider another condition that is more stringent than extremely smooth. A Banach space X is said to be <u>strongly extremely smooth</u> if in X***, $\|x^* + x^\perp\| = \|y^* + y^\perp\| = 1$ and $\|(x^* + x^\perp) + (y^* + y^\perp)\| = 2$ imply that $x^* = y^*$.

In [15] it was shown that if the norm in X is uniformly Gateaux differentiable then X is extremely smooth. It is known that X is uniformly Gateaux differentiable if and only if X* is weak* uniformly rotund, i.e. for all norm-1 sequences (x_k^*) and (y_k^*) in X*, $\|x_k^* + y_k^*\| \to 2$ implies that $x_k^* - y_k^* \to 0$. The following theorem strengthens the result in [15].

<u>Theorem 5</u>. If X is uniformly Gateaux differentiable then X is strongly extremely smooth.

<u>Proof</u>. Suppose that for $\|x^* + x^\perp\| = \|y^* + y^\perp\| = 1$ we have $\|(x^* + x^\perp) + (y^* + y^\perp)\| = 2$ and $(x^* - y^*)(z) \geq a > 0$ for some $\|z\| = 1$. Then using local reflexivity as in [14] and [15] we obtain norm-1 sequences (x_k^*) and (y_k^*) such that $\|x_k^* + y_k^*\| \to 2$ while $(x_k^* - y_k^*)(z) \geq a/2 > 0$. This contradicts the fact that X* is weak* uniformly rotund. QED

The next result gives the relative strength of property W.

<u>Theorem 6</u>. If X is strongly extremely smooth then X has property W.

<u>Proof</u>. Suppose X fails to have property W and let $B_m^n = B(e(n)_m; r(n)_m)$ be a collection of closed balls and (x_k^*), (y_k^*) sequences of functionals satisfying

properties (a) - (d). For convenience we may assume that for all $m \le k \le n$ and all $u \in B_m^n$, $x_k^*(u) \le 2$ and $y_k^*(u) \le 2$. Assume also that for all n, $B_1^n = B(0;1)$, the unit ball.

For all $m \le n$ let $h(n)_m \equiv \frac{-e(n)_m}{r(n)_m - 1}$. From the choice of B_1^n we have that $\|h(n)_m\| \le 1$. For $m \le k \le n$ let

$$\alpha_k(n)_m \equiv \sup y_k^*[B_m^n] \le 2$$

and

$$\beta_k(n)_m \equiv \sup x_k^*[B_m^n] \le 2.$$

Notice that for $m \le k \le n$

$$y_k^*(h(n)_m) = \frac{r(n)_m - \alpha_k(n)_m}{r(n)_m - 1}$$

$$= \frac{1}{1 - \frac{1}{r(n)_m}} - \frac{\alpha_k(n)_m}{r(n)_m}\left(1 - \frac{1}{r(n)_m}\right)$$

so that for m large and $m \le k \le n$, $y_k^*(h(n)_m)$ is arbitrarily close to 1. The same is true for x_k^*.

For each m let H_m in X** be a weak* limit point of the sequence $(h(n)_m)$. If m and k are fixed with $m \le k$ then $H_m(y_k^*)$ is close to $y_k^*(h(n)_m)$ for some $n \ge k$. The previous calculation shows that for m large this is close to 1. The same is true for x_k^*.

Finally, let x*** and y*** be weak* limit points in X*** of (x_k^*) and (y_k^*) respectively. Notice that $\|x^{***}\| \le 1$ and $\|y^{***}\| \le 1$ and from the hypothesis we may assume that $(x^{***} - y^{***})(z) \ge a > 0$ for some $z \in S$. On the other hand, for each m

$$\|x^{***} + y^{***}\| \ge x^{***}(H_m) + y^{***}(H_m) .$$

For some $k \ge m$ this is close to $H_m(x_k^*) + H_m(y_k^*)$ which, for m large, is close to 2. Hence X is not strongly extremely smooth. QED

A rotundity property related to "extremely smooth" was also defined in [15], namely a Banach space X is said to be <u>extremely rotund</u> if every finite dimensional subspace of X is a Chebyshev subspace of X**. It is clear that if X* is extremely smooth then, since X** is rotund, X is extremely rotund. We now introduce a rotundity property related to "strongly extremely smooth". A Banach space X is said to be strongly extremely rotund if for norm-1 vectors x** and y** in X**, $\|x^{**} + y^{**}\| = 2$ implies that $\text{dist}(x^{**} - y^{**}, X) = \|x^{**} - y^{**}\|$. Notice that if X** is rotund then X is strongly extremely rotund. The following sequence of results give a sufficient condition for X to be reflexive in terms of "extremely smooth" and "strongly extremely rotund".

Lemma 7. If X is strongly extremely rotund then X is extremely rotund.

Proof. If X is not extremely rotund then standard arguments from the theory of best approximations yield a nonzero x in X and x^{**} in X^{**} such that $\|x^{**} - x\| = \|x^{**}\| = 1$ and $\|(x^{**} - x) + x^{**}\| = 2$. This contradicts that fact that X is strongly extremely rotund. QED

Recall that $P_1 = J_1 J_0^*$ is the norm-1 projection on X^{***} with range X^* and null space $X^{\perp}$ and that $P_2 = J_2 J_1^*$ is the corresponding projection on $X^{(4)}$ with range X^{**} and null space $X^{*\perp}$. The map P_1 is connected with the result of Dixmier mentioned in the introduction. Brown [2] showed that $\|J_{2_1} - J_0^{**}\| = \|I - P_1\|$. Since $\|P_1\| = 1$ we have $\|I - P_1\| \leq 2$. If $X = c_0$ then $X^{***} = (\ell^1 + c_0{}^{\perp})_{\ell^1}$ and so $\|I - P_1\| = 1$. Brown proved that for $X = \ell^1$, $\|I - P_1\| = 2$. Notice that for any space $\operatorname{dist}(x^*, X^{\perp}) = \|x^*\|$, and $\|I - P_1\| = 1$ if and only if $\operatorname{dist}(x^{\perp}, X^*) = \|x^{\perp}\|$ for all $x^{\perp} \in X^{\perp}$.

Theorem 8. If $\|I - P_1\| = 1$, then X is strongly extremely smooth if and only if X^* is strongly extremely rotund.

Proof. Suppose that for norm-1 vectors x^{***} and y^{***} in X^{***} we have $\|x^{***} + y^{***}\| = 2$. Then assuming X is strongly extremely smooth gives $x^{***} - y^{***} \in X^{\perp}$. Since $\|I - P_1\| = 1$ we have $\operatorname{dist}(x^{***} - y^{***}, X^*) = \|x^{***} - y^{***}\|$ and so X^* is strongly extremely rotund.

Assume now that X^* is strongly extremely rotund and that $x^* + x^{\perp}$ and $y^* + y^{\perp}$ are norm-1 vectors such that $\|(x^* + x^{\perp}) + (y^* + y^{\perp})\| = 2$ and $x^* \neq y^*$. Then using the hypothesis and $\|I - P_1\| = 1$ again gives

$$\begin{aligned} \|x^{\perp} - y^{\perp}\| &\leq \|(x^{\perp} - y^{\perp}) + (x^* - y^*)\| \\ &= \operatorname{dist}((x^{\perp} - y^{\perp}) + (x^* - y^*), X^*) \\ &= \|x^{\perp} - y^{\perp}\|. \end{aligned}$$

Thus both 0 and $-(x^* - y^*)$ are best approximations to $x^{\perp} - y^{\perp}$ in the one dimensional subspace spanned by $x^* - y^*$ in X^{***}. From Lemma 7 this contradicts the assumption that X^* is strongly extremely rotund. QED

Lemma 9. If $\|I - P_2\| = 1$ then $\|I - P_1\| = 1$.

Proof. If $\|I - P_1\| > 1$ then from Brown's theorem there exists $x^{**} \in S^{**}$ such that $\|J_0^{**}x^{**} - J_2 x^{**}\| > 1$. Clearly $J_0^{**}x^{**} - J_2 x^{**} \in X^{*\perp}$. However, $\|J_2 x^{**} + (J_0^{**}x^{**} - J_2 x^{**})\| = 1$ which contradicts $\|I - P_2\| = 1$. QED

<u>Theorem 10</u>. If $\|I - P_2\| = 1$ and X^{**} is extremely smooth then X is reflexive.

<u>Proof</u>. First notice that for any space Y if Y^* is extremely smooth and $\|I - P_1\| = 1$ then $\text{dist}(y^{**}, Y) < 1$ for all $\|y^{**}\| = 1$. In fact, if $\text{dist}(y^{**}, Y) = 1$ then $y^{\perp}(y^{**}) = 1$ for some $\|y^{\perp}\| = 1$. However, $\|I - P_1\| = 1$ gives that $\text{dist}(y^{\perp}, Y^*) = \|y^{\perp}\|$, and from the characterization given in Theorem 1 applied to Y^* and Y^{***} we have a contradiction.

Now, X^{**} extremely smooth and $\|I - P_2\| = 1$ says that $\text{dist}(x^{***},X^*) < 1$ for all $\|x^{***}\| = 1$. However, from the previous lemma $\|I - P_1\| = 1$ and so for all $\|x^{\perp}\| = 1$ we have $\text{dist}(x^{\perp},X^*) = 1$. QED

<u>Lemma 11</u>. If X^{**} is strongly extremely rotund then $\|I - P_1\| = 1$.

<u>Proof</u>. If $\|I - P_1\| > 1$ then from Brown's result there exists $x^{**} \in S^{**}$ such that $\|J_2x^{**} - J_0^{**}x^{**}\| > 1$ and $\|J_2x^{**} + J_0^{**}x^{**}\| = 2$. However, it is clear that

$$\text{dist}(J_2x^{**} - J_0^{**}x^{**},\ X^{**}) \leq 1.$$

This contradicts the fact that X^{**} is strongly extremely rotund. QED

<u>Theorem 12</u>. If X^{***} is strongly extremely rotund and X^{**} is extremely smooth then X is reflexive.

<u>Proof</u>. From Lemma 11 we have $\|I - P_2\| = 1$. Now the result follows from Theorem 10. QED

<u>Corollary</u>. If X^{***} is strongly extremely rotund and $\|I - P_3\| = 1$ where $P_3 = J_3J_2^*$, then X is reflexive.

<u>Proof</u>. Use Theorem 8 and Theorem 12. QED

We now summarize in a chart the implications that exist among the various geometrical properties that we have considered. We follow the notation used in [15]. If P is a geometrical property then P^k means that the kth dual of X has P. We write simply P for P^0. The usual arrow is used to indicate implication. The symbol + beside an arrow indicates the additional hypothesis that $\|I - P_1\| = 1$. We employ the following abbreviations:

R - rotund (strictly convex)
S - smooth
ER - extremely rotund
ES - extremely smooth
SER - strongly extremely rotund
SES - strongly extremely smooth

w*UR - weak* uniformly rotund

UG - uniformly Gateaux differentiable

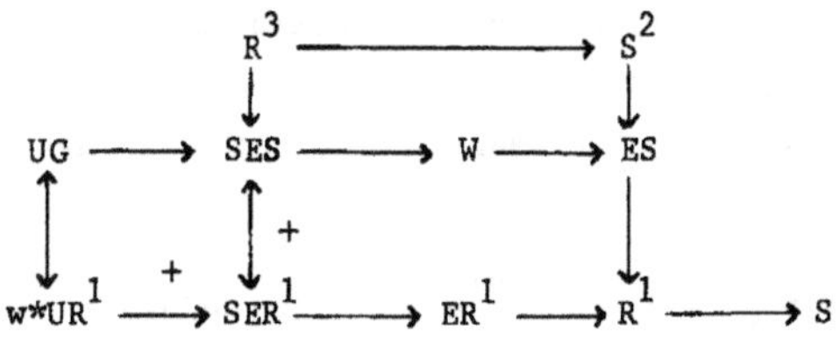

Recall from [15] that a Banach space X is said to be <u>Hahn-Banach smooth</u> if in X^{***}, $\|x^* + x^\perp\| = \|x^*\| = 1$ implies that $x^\perp = 0$. In other words $x^* \in X^{***}$ is the unique Hahn-Banach extension of $x^*|_X$. The following is an improvement of the result in [14].

<u>Theorem 13</u>. If $\dim(X^{**}/X) \leq 1$, then X is strongly extremely smooth and Hahn-Banach smooth if and only if X^{***} is rotund.

<u>Proof</u>. If X^{***} is rotund then it is clear that X is strongly extremely smooth and Hahn-Banach smooth.

For the converse we follow the scheme in [14]. Suppose $\|(x^* + ax^\perp) + (y^* + bx^\perp)\| = 2$ for norm-1 vectors $x^* + ax^\perp$ and $y^* + bx^\perp$ in $X^{***} = X^* \oplus [x^\perp]$ where $[x^\perp]$ denotes the span of $x^\perp$. Since X is strongly extremely smooth we have $x^* = y^*$. Now choose $\|x^{(4)}\| = 1$ such that

$$x^{(4)}(x^* + ax^\perp) = x^{(4)}(x^* + bx^\perp) = 1.$$

Then $(a - b)x^{(4)}(x^\perp) = 0$. If $x^{(4)}(x^\perp) \neq 0$ then $a = b$ and the proof is complete. Otherwise $x^{(4)}(x^\perp) = 0$ and hence $\|x^*\| = 1$. But now since X is Hahn-Banach smooth we have $a = 0$. Similarly it follows that $b = 0$. QED

In [15] it was shown that if X is Hahn-Banach smooth and smooth then X is very smooth and consequently X^* has the Radon-Nikodym property. In [10] Namioka and Phelps considered the property (**) in dual spaces (X^* has property (**) if for every net (x^*_α) in X^*, $(x^*_\alpha) \overset{*}{\to} x^*$ and $\|x^*_\alpha\| \to \|x^*\|$ imply that $x^*_\alpha \to x^*$.) and showed that if X^* has property (**) then X is an Asplund space and hence X^* has the Radon-Nikodym property. It is also known that if X^* has property (**) then X is Hahn-Banach smooth [15].

We now introduce a property less stringent than Hahn-Banach smooth which will allow us to improve both the results mentioned above. A Banach space X is said to be <u>weakly Hahn-Banach smooth</u> if in X^{***}, $\|x^* + x^\perp\| = \|x^*\| = 1$ and $x^*(x) = \|x\| = 1$ imply that $x^\perp = 0$. In other words for each $x^* \in S^*$ that attains its norm on S, $x^* \in X^{***}$ is the unique Hahn-Banach extension of $x^*|_X$.

Lemma 14. X is very smooth if and only if X is smooth and weakly Hahn-Banach smooth.

Proof. This follows immediately from the definitions of these properties. QED

Theorem 15. If X is weakly Hahn-Banach smooth then X* has the Radon-Nikodym property.

Proof. Let Y be a separable subspace of X. Using the Hahn-Banach Theorem and noticing that the extension of $y^* + y^{\perp}$ is not weak* continuous for $y^{\perp}$ nonzero we have that Y is also weakly Hahn-Banach smooth. Now, using Uhl's criterion for dual spaces to possess the Radon-Nikodym property it suffices to show that Y* is separable [18].

Assuming to the contrary we follow the scheme given in [12]. Let $\{y_n\}$ be a dense set in the unit sphere of Y. Choose $\|y_n^*\| = 1$ such that $y_n^*(y_n) = 1$ and let H be the subspace of Y* generated by $\{y_n^*\}$. Notice that $\|y\| = \sup\{|h(y)| : h \in H\}$ for all $y \in Y$. Since $H \neq Y^*$ and H is norm closed, using the Bishop-Phelps Theorem we can find $\|z^*\| = 1$ in $Y^* \setminus H$ and $\|z\| = 1$ in Y such that $z^*(z) = 1$. Choose $\|z^{**}\| = 1$ in $H^{\perp} \subseteq Y^{**}$ such that $z^{**}(z^*) \neq 0$. Now, define f on $Y \oplus H^{\perp} \subseteq Y^{**}$ by $f(y + h^{\perp}) = z^*(y)$. Notice that $\|f\| = 1$ since $\|y\| \leq \|y + h^{\perp}\|$ for all $y \in Y$ and $h^{\perp} \in H^{\perp}$. Since f is an extension of z* to $Y \oplus H^{\perp}$, by the Hahn-Banach Theorem we can extend f to a norm-1 functional on all of Y**; call this extension $z^* + z^{\perp}$. Then we have $\|z^* + z^{\perp}\| = \|z^*\| = 1$ and $z^*(z) = \|z\| = 1$ while $(z^* + z^{\perp})(z^{**}) = f(z^{**}) = 0$, since $z^{**} \in H^{\perp}$, and hence $z^{\perp}(z^{**}) = -z^{**}(z^*) \neq 0$. This contradicts the fact that Y is weakly Hahn-Banach smooth. QED

REFERENCES

1. E. Bishop and R. R. Phelps, The support functionals of a convex set, Convexity, 27-35 (Proc. Sympos. Pure Math. 7 Amer. Math. Soc., Providence, Rhode Island, 1963).
2. A. L. Brown, On the canonical projection of the third dual of a Banach space onto the first dual, (to appear).
3. M. M. Day, Strict convexity and smoothness of normed spaces, Trans. Amer. Math. Soc. 78(1955), 516-528.
4. ________, Normed linear spaces, 3rd ed. (Ergebnisse der Mathematik und ihrer Grenzgebiete, 21. Springer-Verlag, Berlin, Heidelberg, New York, 1973).
5. D. W. Dean, The equation L(E,X**) = L(E,X)** and the principle of local reflexivity, Proc. Amer. Math. Soc. 40(1973), 146-148.
6. J. Diestel and B. Faires, On vector measures, Trans. Amer. Math. Soc. 198(1974), 253-271.
7. J. Dixmier, Sur un theoreme de Banach, Duke Math. J. 15(1948), 1057-1071.
8. J. R. Giles, On smoothness of the Banach space embedding, Bull. Austral. Math. Soc. 13(1975), 69-74.
9. J. Lindenstrauss and H. P. Rosenthal, The $\mathcal{L}_p$ spaces, Israel J. Math. 7(1969), 325-349.
10. I. Namioka and R. R. Phelps, Banach spaces which are Asplund spaces, Duke Math. J. 42(1975), 735-750.
11. J. Rainwater, A non-reflexive Banach space has non-smooth third conjugate space, unpublished note.

12. I. Singer, On the problem of non-smoothness of non-reflexive second conjugate spaces, Bull. Austral. Math. Soc. 12(1975), 407-416.
13. M. A. Smith, A smooth, non-reflexive second conjugate space, Bull. Austral. Math. Soc. 15(1976), 129-131.
14. ________, Rotundity and smoothness in conjugate spaces, (to appear).
15. F. Sullivan, Geometrical properties determined by the higher duals of a Banach space, (to appear).
16. D. G. Tacon, The conjugate of a smooth Banach space, Bull. Austral. Math. Soc. 2(1970), 415-425.
17. S. Troyanski, Example of a smooth space whose conjugate has not strictly convex norm, Studia Math. 35(1970), 305-309.
18. J. J. Uhl, Jr., A note on the Radon-Nikodym property, Revue Roumaine de Math. Pures et Appl. 17(1972), 113-115.
19. L. P. Vlasov, Approximative properties of sets in normed linear spaces, Russian Math. Surveys 28(1973), 1-66.

A PROOF OF THE MARTINGALE CONVERGENCE THEOREM IN BANACH SPACES

Charles Stegall
Kaiserstrasse 43
522 Waldbröl
Bundesrepublik Deutschland

We give here an elementary proof of the vector valued martingale convergence theorem. Specifically, we prove the following theorem first proved by Chatterji [1]:

Theorem: Let X be a Banach space with the Radon-Nikodym property. If $(\overline{M}_n)$ is an X-valued, bounded martingale, then $(\overline{M}_n)$ converges a. s. (converges in probability if the martingale is not discrete).

We point out that the converse of this theorem is also true and very easy to prove.

Our proof is elementary in the sense that we use only basic facts from functional analysis and the theorem of Doob (see the appendix of [3] for an elementary proof) that discrete, bounded, real valued martingales converge a. s. (converge in probability in the non discrete case, as was proved by Krickeberg: see [4]). We do not use any results concerning sub- or supermartingales nor any results about finitely additive set functions.

Let (S, Σ, P) be a probability space and X a Banach space. An X-valued martingale (denoted simply by $(\overline{M}_n)$) is a directed set N, a collection of P-complete sigma-algebras Σ_n indexed by N, with Σ_n contained in Σ_m if $n \leq m$ and Bochner integrable functions $\overline{M}_n: S \to X$ with $E(\overline{M}_m, \Sigma_n) = \overline{M}_n$ if $n \leq m$ (E denotes the conditional expectation operator). A function $\overline{M}_n: S \to X$ is Bochner integrable if it is Borel measurable, essentially separably valued and

$$\int \|\overline{M}\| \, dP < +\infty.$$

We say that a martingale is bounded if

$$\sup_n \int \|\overline{M}_n\| \, dP < +\infty.$$

We always assume Σ is the sigma-algebra generated by $\cup\Sigma_n$. A Banach space has the Radon-Nikodym property (RNP) if for any probability space (S, Σ, P), any $\overline{m}: \Sigma \to X$ that is countably additive, of finite variation, absolutely continuous with respect to P, is representable by a Bochner integrable function. That is, there exists an $\overline{M}: S \to X$ such that

$$\overline{m}(C) = \int_C \overline{M} \, dP \text{ for all } C \text{ in } \Sigma \text{ (see [2]).}$$

By $\bar{M}_n \to \bar{M}$ we mean that $\bar{M}_n$ converges to $\bar{M}$ in probability (converges a. s. in the discrete case). A Banach space X is always considered as a canonical subspace of X^{**}.

We need the following elementary fact: if $\bar{M}: S \to X$ is Bochner integrable then $E(\bar{M}, \Sigma_n) \to \bar{M}$.

The following lemma is known (see [4]):

<u>Lemma</u>: Let $(\bar{M}_n)$ be a bounded X-valued martingale such that for each x^* in X^* we have that $x^*\bar{M}_n \to 0$. Then $\|\bar{M}_n\| \to 0$.

Proof: Let $M_n = \sup_{n \leq m} E(\|\bar{M}_m\|, \Sigma_n)$. Then (M_n) is a positive martingale, and if we assume that

$$\sup_n \int \|\bar{M}_n\| \, dP = 1,$$

there exists M such that $M_n \to M$. Given $\epsilon > 0$, there exists n' such that if $n \geq n'$ then

$$\int_S \|\bar{M}_n\| \, dP > 1 - \epsilon/2.$$

Since $\bar{M}_{n'}$ has an essentially separable range there exists a sequence (x^*_k) in X^*, $\|x^*_k\| = 1$, such that

$$\|\bar{M}_{n'}(s)\| = \sup_k |x^*_k \bar{M}_{n'}(s)| \text{ a. s.}$$

Define $h_{n,k}(s) = \sup_{i \leq k} |x^*_i \bar{M}_n(s)|$ a. s. Then

(i) $h_{n,k} \leq h_{n,k+1}$ a. s.

(ii) $h_{n,k} \leq E(h_{m,k}, \Sigma_n)$ a. s. if $n \leq m$.

By (i) and the dominated convergence theorem there exists m' such that for $m' \leq m$, $\int h_{n',m} \, dP \geq 1 - \epsilon$. By hypothesis, $h_{n,m'} \to 0$; thus, $M_n - h_{n,m'} \to M$, and by Fatou's lemma

$$0 \leq \int_S M \, dP \leq \liminf \left(\int_S (M_n - h_{n,m'}) \, dP = \liminf \left(1 - \int_S h_{n,m'} \, dP\right) \leq \epsilon.$$

Thus $M = 0$ a. s. and since $\|\bar{M}_n\| \leq M_n$ we have that $\|\bar{M}_n\| \to 0$.

<u>Proposition</u>: Let $(\bar{M}_n)$ be a bounded X-valued martingale. Let

$$R: X^* \to L_1(S, \Sigma, P)$$

be the unique, continuous, linear function such that

$$Rx^* = \lim_n x^* \bar{M}_n.$$

Then R^* is weakly compact and $R^*(L_\infty(S, \Sigma, P))$ is a subset of X.

Proof: Assume $\sup_n \int_S \|\overline{M}_n\| \, dP = 1$. The function R exists by Doob's theorem, is clearly linear, and by Fatou's lemma, $\|R\| \leq 1$. As before, if $M_n = \sup_{n \leq m} E(\|\overline{M}_m\|, \Sigma_n)$ is a positive martingale and $\int_S M_n \, dP = 1$. Let H_0 be the vector subspace of $L_\infty(S, \Sigma, P)$ such that each h in H_0 is Σ_n measurable for some n. For h and g in H_0, define

$$\langle f, g \rangle = \lim_n \int_S fgM_n \, dP.$$

If $K = (f\colon \langle f, g\rangle = 0$ for all g in $H_0)$ and H is the quotient space H_0/K, then $(H, \langle\, , \rangle)$ is a pre-Hilbert space. Define

$$T\colon H \to X \text{ by } Th = \lim_n \int_S h\overline{M}_n \, dP.$$

Then, if h is Σ_n measurable

$$\|Th\| \leq \int_S |h| \, \|\overline{M}_n\| \, dP \leq \int_S |h| \, M_n \, dP \leq \left(\int_S h^2 M_n \, dP\right)^{1/2} \left(\int_S M_n \, dP\right)^{1/2}.$$

This shows that T is a continuous linear operator, $\|T\| \leq 1$ and if W denotes the smallest closed, convex subset of X containing $(Th\colon \langle h, h\rangle \leq 1)$ then W is a weakly compact set. Suppose that for some g in $L_\infty(S, \Sigma, P)$, $\|g\| \leq 1$, R^*g is not in W. Then there exists an x^* in X^* and a real number $B \geq 0$ such that

$$(R^*g)(x^*) > B \geq x^* \, Th \text{ for all h in H, } \langle h, h\rangle \leq 1.$$

Let $h_n = \operatorname{sgn} x^* \overline{M}_n$. Then h_n is in H and $\langle h_n, h_n\rangle \leq 1$. Thus,

$$B \geq x^*(Th_n) = x^*\left(\int_S h_n \overline{M}_n \, dP\right) = \int_S h_n (x^* \overline{M}_n) \, dP = \int_S |x^* \overline{M}_n| \, dP,$$

and, by Fatou's lemma, $\int_S |Rx| \, dP \leq B$. However,

$$B < \int_S (Rx^*) g \, dP \leq \int_S |Rx| \, dP.$$

This is a contradiction.

We point out that the proof of the proposition did not require that X have RNP.

To prove the theorem, let (using the same notation as in the proposition) $\overline{m}(C) = R^* I_C$ (where I_C denotes the characteristic function of the set C in Σ). Suppose the real valued martingale (M_n) converges (in the appropriate sense) to M. Then, for C in Σ and $\|x^*\| = 1$,

$$I_C \, |x^* \overline{M}_n| \to I_C \, |Rx^*| \text{ and}$$

$$I_C\,|x^*\,\overline{M}_n| \leq I_C\,\|\overline{M}_n\| \leq I_C M_n \text{ a. s. and}$$

$$I_C M_n \to I_C\,M.$$

Therefore, $|Rx^*| \leq M$ a. s. for all $\|x^*\| \leq 1$ and

$$\|R^*I_C\| \leq \int_C M\,dP \text{ for all } C \text{ in } \Sigma.$$

Thus, $\overline{m}$ is represented by the Bochner integrable function $\overline{M}_\infty$ and

$$x^*(\int_C \overline{M}_\infty\,dP) = x^*(R^*I_C) = \int_C Rx^*\,dP = \int_C (\lim_n x^*\overline{M}_n)\,dP.$$

This proves that for all x^* in X^*

$$x^*\,\overline{M}_n \to x^*\,\overline{M}_\infty \text{ and } x^*(\overline{M}_n - E(\overline{M}_\infty, \Sigma_n)) \to 0.$$

Combining the lemma with the fact that $E(\overline{M}_\infty, \Sigma_n) \to \overline{M}_\infty$ we have $\overline{M}_n \to \overline{M}$.

REFERENCES

1. S. D. Chatterji, Martingale convergence and the Radon-Nikodym theorem in Banach spaces, Math. Scand. 22(1968), 21-41.
2. J. Diestel and J. J. Uhl, Jr., The Theory of Vector Measures, to appear in the American Mathematical Society Surveys.
3. P. A. Meyer, Martingales and Stochastic Integrals, Lecture Notes in Mathematics, Vol. 284, Springer-Verlag, Berlin 1972.
4. J. Neveu, Martingales a temps discret, Masson and Cie, Paris 1975.

ACKNOWLEDGEMENT: The author would like to thank Richard Haydon for the discussions he had with him about the results of this paper. This paper was prepared while the author was a member of the Sonderforschungsbereich 72, Institut für Angewandte Mathematik, Universität Bonn.